6th IF-YITP GR+HEP+COSMO International Symposium

Journal of Physics: Conference Series Volume 883

Phitsanulok, Thailand
3 – 5 August 2016

ISBN: 978-1-5108-4723-1
ISSN: 1742-6588

Printed from e-media with permission by:

Curran Associates, Inc.
57 Morehouse Lane
Red Hook, NY 12571

Some format issues inherent in the e-media version may also appear in this print version.

Printed by Curran Associates, Inc. (2017)

For permission requests, please contact the Institute of Physics
at the address below.

Institute of Physics
Dirac House, Temple Back
Bristol BS1 6BE UK

Phone: 44 1 17 929 7481
Fax: 44 1 17 920 0979

techtracking@iop.org

Additional copies of this publication are available from:

Curran Associates, Inc.
57 Morehouse Lane
Red Hook, NY 12571 USA
Phone: 845-758-0400
Fax: 845-758-2633
Email: curran@proceedings.com
Web: www.proceedings.com

Table of contents

Volume 883

IF-YITP GR+HEP+Cosmo International Symposium VI
3–5 August 2016, Naresuan University, Thailand

Accepted papers received: 18 August 2017
Published online: 24 August 2017

IOP Conf. Series: Journal of Physics: Conf. Series **883** (2017) 011001 doi:10.1088/1742-6596/883/1/011001

Preface: IF+YITP GR+HEP+Cosmo International Symposium VI

Piero Nicolini

E-mail: nicolini@fias.uni-frankfurt.de
Frankfurt Institute for Advandced Studies (FIAS), Frankfurt am Main, Germany
and
Institut für Theoretische Physik, Goethe-Universitat Frankfurt am Main, Germany

Douglas Singleton

E-mail: dougs@mail.fresnostate.edu
Department of Physics, California State University Fresno, CA, USA
and
Institute of Experimental and Theoretical Physics Al-Farabi KazNU, Almaty, Kazakhstan

Shingo Takeuchi

E-mail: shingo.portable@gmail.com
The Institute for Fundamental Study, "The Tah Poe Academia Institute", Naresuan University, Phitsanulok, Thailand

Matthew J. Lake

E-mail: matthewj@nu.ac.th
The Institute for Fundamental Study, "The Tah Poe Academia Institute", Naresuan University, Phitsanulok, Thailand
and
Thailand Center of Excellence in Physics, Ministry of Education, Bangkok, Thailand

Vladimir Dzhunushaliev

E-mail: v.dzhunushaliev@gmail.com
Institute of Experimental and Theoretical Physics Al-Farabi KazNU, Almaty, Kazakhstan

INTRODUCTION

The Institute for Fundamental Study (IF) hosted the 6th IF+YITP International Symposium on superstring theory, the quantum theory of fields, gravity, cosmology, and related topics, on August 3rd through 5th, 2016, on the campus of Naresuan University, Thailand. This version of the symposium was co-sponsored by the Yukawa Institute for Theoretical Physics (YITP). These symposia have two principle aims. First, to bring together researchers from Thailand and around the world to discuss new ideas in fundamental physics. Second, these meetings are to support the recently established Institute for Fundamental Study (The Tah Poe Academia

IOP Conf. Series: Journal of Physics: Conf. Series **883** (2017) 011001 doi:10.1088/1742-6596/883/1/011001

Institute) centered in Naresuan University in Phitsanulok, Thailand. The 6th symposium was a great success in terms of these two goals. The meeting had eight invited speakers and over 60 participants from both inside and outside Thailand. A host of new models and theories were presented at this symposium which covered a wide range of theoretical topics from cosmology to quantum field theory to superstring theory. The list of invited speakers and their talk titles is given below.

- Alexei A. Starobinsky (Landau Inst., Russia): "Inflaton potential: smooth global reconstruction and fine features"
- Chethan Krishnan (Indian Inst. Sci., India): "Neumann gravity" and "Higher gravity"
- Edouard B. Manoukian (IF, Thailand): "Developments of Some Fundamental Problems in Quantum Field Theory and Quantum Physics"
- Feng-Li Lin (National Taiwan Normal U): "Relative entropy and torsion coupling" and "Anti-Unruh phenomena in the open dynamics of Majorana zero modes"
- Kazuki Hasebe (Sendai National College of Technology, Japan): "Nambu geometry in quantum Hall effect and topological insulators"
- Misao Sasaki (YITP): "Inflationary Magnetogenesis with Broken Local U(1) Symmetry"
- Seiji Terashima (YITP): "Exact Path Integral for 3D Quantum Gravity"
- Shinji Mukohyama (YITP): "Is DBI as fragile as other k-essence fields? and "Ghost condensation and horizon entropy"

A more complete list of participants as well as a schedule of talks can be found at
http://www.if.nu.ac.th/node/1407

and

http://www.if.nu.ac.th/sites/default/files/bin/talk_schedule13.pdf respectively.

The general conference website is

http://www.if.nu.ac.th/IF_YITP_GR_HEP_Cosmo_International_Symposium_VI .

COMMITTEES

The various committees who helped with the scientific and logistical organization of this meeting were

- **International Advisory Commitee**
  Emil T. Akhmedov (ITEP)
  Misao Sasaki (YITP)
  Douglas Singleton (California State U)
  Shinji Tsujikawa (Tokyo University of Science)
- **Scientific Organizing Commitee**
  Antonio De Felice (YITP)
  Burin Gumjudpai (IF, Naresuan U)
  Edouard B. Manoukian (IF, Naresuan U)
  Seckson Sukhasena (IF, Naresuan U)
- **Local Organizers** Shingo Takeuchi (IF, Naresuan U)
  Matthew J. Lake (IF, Naresuan U)

IOP Conf. Series: Journal of Physics: Conf. Series **883** (2017) 011001 doi:10.1088/1742-6596/883/1/011001

LIST OF PARTICIPANTS

Alexei A. Starobinsky (Russia): Landau Inst. (Russia)
Antonio De Felice (Italy): YITP (Japan)
Atirat Maksuwan (Thailand): Pathumwan Inst. of Technology (Thailand)
Auttakit Chatrabhuti (Thailand): Chulalongkorn U (Thailand)
Atri Deshamukhya (India): Assam U (India)
Chaiyapoj Muthaporn (Thailand): Khon Kaen university (Thailand)
Chanadan Douykumklaw (Thailand): U of Phayao (Thailand)
Chethan Krishnan (India): Indian Inst. Sci. (India)
Feng-Li Lin (Taiwan): National Taiwan Normal U (Taiwan)
Hai-Qing Zhang (China): Utrecht U (The Netherlands) (CANCELED)
Hiroshi Isono (Japan): Chulalongkorn U (Thailand)
Kazuki Hasebe (Japan): Sendai National College of Technology (Japan)
Kittipong Aimsamer (Thailand): King Mongkuts U of Tech. Thonburi (Thailand)
Kunchadapa Tanatip (Thailand): U of Phayao (Thailand)
Maneenate Wechakama (Thailand): Kasetsart U (Thailand)
Nakorn Thongyoi (Thailand): Khon Kaen university (Thailand)
N Malsawmtluangi (India): University of Hyderabad (India)
Misao Sasaki (Japan): YITP (Japan)
Pallab Basu (India): International Center for Theoretical Sciences (India)
Phongsaphat Rangdee (Thailand): U of Phayao (Thailand)
Pisuttawan Sripirom Sirininlakul (Thailand): Srinakharinwirot U (Thailand)
Ryotaro Kase (Japan): Tokyo U of Science (Japan)
Sachiko Kuroyanagi (Japan): Nagoya U (Japan)
Samak Boonpan (Thailand): Khon Kaen university (Thailand)
Seiji Terashima (Japan): YITP (Japan)
Siri Sirininlakul (Thailand): Srinakharinwirot U (Thailand)
Shin Sasaki (Japan): Kitazato U (Japan)
Shinji Mukohyama (Japan): YITP (Japan)
Shinji Tsujikawa (Japan): Tokyo U of Science (Japan)
Sikarin Yoo-Kong (Thailand): King Mongkuts U of Tech. Thonburi (Thailand)
Sirachak Panpanich (Thailand): Waseda U (Japan)
Sudhakar Panda (India): Institute of Physics (India)
Yuki Sato (Japan): Chulalongkorn U (Thailand)
Watcharanon Kantayasakun (Thailand): King Mongkuts U of Tech. Thonburi (Thailand)

IF members

Anajak Phonchantuek (Thailand): IF, Naresuan U (Thailand)
Burin Gumjudpai (Thailand): IF, Naresuan U (Thailand)
Chomphet Maruekhen (Thailand): IF, Naresuan U (Thailand)
Chonticha Kritpetch (Thailand): IF, Naresuan U (Thailand)
Edouard B. Manoukian (Canada): IF, Naresuan U (Thailand)
Jarunee Niemtong (Thailand): IF, Naresuan U (Thailand)
Jorge Ovalle (Chile and Venezuela): IF, Naresuan U (Thailand) and Simon Bolivar U (Venezuela)
Khamphee Karwan (Thailand): IF, Naresuan U (Thailand)
Lunchakorn Tannukij (Thailand): IF, Naresuan U (Thailand)
Matthew J. Lake (UK): IF, Naresuan U (Thailand)
Narakorn Kaewkhao (Thailand): IF, Naresuan U (Thailand)
Nopadhol Kamma (Thailand): IF, Naresuan U (Thailand)

Paichayon Tongsin (Thailand): IF, Naresuan U (Thailand)
Pichet Vanichchapongjaroen (Thailand): IF, Naresuan U (Thailand)
Pitayuth Wongjun (Thailand): IF, Naresuan U (Thailand)
Rakpong Saikaew (Thailand): IF, Naresuan U (Thailand)
Ratchaphat Nakarachinda (Thailand): IF, Naresuan U (Thailand)
Sakdithut Jitpienka (Thailand): IF, Naresuan U (Thailand)
Seckson Sukhasena (Thailand): IF, Naresuan U (Thailand)
Sheng-Lan Ko (Taiwan): IF, Naresuan U (Thailand)
Shingo Takeuchi (Japan): IF, Naresuan U (Thailand)
Somphoach Saichaemchan (Thailand): IF, Naresuan U (Thailand)
Stharporm Sapa (Thailand): IF, Naresuan U (Thailand)
Sujiphat Janaun (Thailand): IF, Naresuan U (Thailand)
Teeraparb Chantavat (Thailand): IF, Naresuan U (Thailand)
Wittaya Thipaksorn (Thailand): IF, Naresuan U (Thailand)
Yuttana Jawralee (Thailand): IF, Naresuan U (Thailand)

Acknowledgments

The meeting was co-sponsored by the Asia Pacific Center for Theoretical Physics (APCTP), Pohang, Korea, Naresuan University, Phitsanulok, Thailand and the Yukawa Institute for Theoretical Physics (YITP), Kyoto, Japan. IF thanks the APCTP for providing generous financial support for several of the workshop participants and the YITP for their financial and logistical support for the meeting.

TAH POE THE INSTITUTE IF THEORETICAL PHYSICS for FUNDAMENTAL STUDY

ASIA PACIFIC CENTER FOR THEORETICAL PHYSICS APCTP

YITP YUKAWA INSTITUTE FOR THEORETICAL PHYSICS

IOP Conf. Series: Journal of Physics: Conf. Series **883** (2017) 011002 doi:10.1088/1742-6596/883/1/011002

Peer review statement

All papers published in this volume of *Journal of Physics: Conference Series* have been peer reviewed through processes administered by the proceedings Editors. Reviews were conducted by expert referees to the professional and scientific standards expected of a proceedings journal published by IOP Publishing.

IOP Conf. Series: Journal of Physics: Conf. Series **883** (2017) 012001 doi:10.1088/1742-6596/883/1/012001

Is there a connection between "dark" and "light" physics?

Matthew J. Lake

The Institute for Fundamental Study, "The Tah Poe Academia Institute", Naresuan University, Phitsanulok 65000, Thailand, and
Thailand Center of Excellence in Physics (ThEP), Ministry of Education, Bangkok 10400, Thailand

E-mail: matthewj@nu.ac.th

Abstract. In the early-mid 20th century Dirac and Zel'dovich were among the first scientists to suggest an intimate connection between cosmology and atomic physics. Though a revolutionary proposal for its time, Dirac's Large Number Hypothesis (1937) adopted a standard assumption of the day, namely, the non-existence of the cosmological constant term ($\Lambda = 0$). As a result, its implementation necessitated extreme violence to the theory of general relativity – something few physicists were prepared to sacrifice in favour of 'numerology' – requiring a time-dependent gravitational coupling of the form $G(t) \sim 1/t$. Zel'dovich's insight (1968) was to realise that a small but nonzero cosmological term ($\Lambda > 0$) allowed the present day radius of the Universe to be identified with the de Sitter radius, $r_U \simeq l_{dS} \simeq 1/\sqrt{\Lambda}$, which removed the need for time-dependence in the fundamental couplings. Thus, he obtained the formula $\Lambda \simeq m^6 G^2/\hbar$, where m is a mass scale characterising the relative strengths of the gravitational and electromagnetic interactions, which he identified with the proton mass m_p. In this paper, we review a number of recent arguments which, instead, suggest the identification $m = m_e/\alpha_e$, where m_e is the electron mass and $\alpha_e = e^2/\hbar c \simeq 1/137$ is the usual fine structure constant. We note that these are of a *physical* nature and, therefore, represent an attempt to lift previous arguments *à la* Dirac from the realm of numerology into the realm of empirical science. If valid, such arguments suggest an intimate connection, not only between the macroscopic and microscopic worlds, but, perhaps even more surprisingly, between the very essence of "dark" and "light" physics.

1. Introduction – Dirac, Zel'dovich and the Large Number Hypothesis

In 1937, Dirac noted the approximate order of magnitude equivalence between several large dimensionless numbers obtained from atomic physics and cosmology [1]. These included the ratio of the present day radius of the Universe, $r_U \simeq 4.40 \times 10^{28}$cm, to the classical electron radius, $r_e = 2.818 \times 10^{-13}$cm, and the ratio of the electric and gravitational forces between an electron and a proton,

$$\frac{r_U}{r_e} \simeq 10^{40} \cong \frac{e^2}{G m_e m_p} \simeq 10^{39} , \tag{1}$$

where $m_e = 9.109 \times 10^{-28}$g and $m_p = 1.672 \times 10^{-24}$g. (For convenience, from here on, we use the most recent measured, or inferred, values of physical length and mass scales.) Interpreting this as a signature of an as yet unknown connection between cosmological and atomic physics,

IOP Conf. Series: Journal of Physics: Conf. Series **883** (2017) 012001 doi:10.1088/1742-6596/883/1/012001

he formulated the Large Number Hypothesis (LNH), which states that the numerical equality between two very large quantities with similar physical meanings cannot be a simple coincidence [1, 2, 3, 4]. However, since the radius of the Universe is not constant but scales approximately as $r_U(t) \sim t$, Dirac surmised that the gravitational coupling varies according to $G(t) \sim 1/t$ – under the assumption $\Lambda = 0$ – though this is, of course, not the only way to satisfy Eq. (1) for all epochs [5]. Nonetheless, such variation is strongly at odds with the current 'concordance' model of cosmology, which is based on Einstein's theory with a positive cosmological term and an unknown "dark" matter component [6].

In 1968, Zeldovich noted the same approximate equivalence between the ratio of the radius of the Universe and the Compton wavelength of the proton, $\lambda_p = h/m_p c = 1.321 \times 10^{-13}$cm, and between λ_p and the proton's Schwarzschild radius, $r_S(m_p) = 2Gm_p/c^2 = 2.484 \times 10^{-52}$cm [7]. In addition, he noted that, if $\Lambda > 0$ and $r_U \simeq l_{dS} \simeq 1/\sqrt{\Lambda}$ – contrary to Dirac's original assumptions – then

$$\frac{r_U}{\lambda_p} \simeq \frac{m_p c}{h\sqrt{\Lambda}} \simeq 10^{40} \cong \frac{\lambda_p}{r_S(m_p)} \simeq \frac{hc}{Gm_p^2} \simeq 10^{39}\,, \tag{2}$$

and hence

$$\Lambda \simeq \frac{m_p^6 G^2}{h^4} \simeq 10^{-56}\text{cm}^{-2}\,. \tag{3}$$

Setting $\Lambda \simeq 1/l_{dS}^2$, where l_{dS} is the de Sitter horizon, this is equivalent to

$$\lambda_p = \frac{h}{m_p c} \simeq (r_{Pl}^2 l_{dS})^{1/3}\,, \tag{4}$$

with $l_{dS} \simeq 10^{28}$cm, where $r_{Pl} \equiv \sqrt{hG/c^3}$ is the Planck length. Zel'dovich's formula (3) is remarkable in that, not only does it establish a possible connection between dark energy (Λ), canonical gravity (G) and elementary particle physics (m_p), it is *also* compatible with current experimental bounds on the dark energy density [8, 9].

The current best fit to the available cosmological data favours a cosmological concordance, or ΛCDM model, in which dark energy takes the form of a positive cosmological constant, which accounts for approximately 69% of the total energy density of the Universe, whereas cold dark matter (CDM) accounts for around 26% and ordinary visible matter for around 5% [8, 9]. The present day density is close to the critical density,

$$\rho_{crit} = \frac{3\mathcal{H}_0^2}{8\pi G} = 8.639 \times 10^{-30}\text{gcm}^{-3}\,, \tag{5}$$

where $\mathcal{H}_0 = 67.74\text{kms}^{-1}\text{Mpc}^{-1} = 2.198 \times 10^{-18}\ \text{s}^{-1}$ is the Hubble constant, so that the dark energy density is

$$\rho_\Lambda = \frac{\Lambda c^2}{8\pi G} = 5.971 \times 10^{-30}\text{gcm}^{-3}\,. \tag{6}$$

This yields an inferred value of the cosmological constant,

$$\Lambda = 1.114 \times 10^{-56}\text{cm}^{-2}\,, \tag{7}$$

which corresponds to a de Sitter radius, $l_{dS} = 1.641 \times 10^{28}$cm.

IOP Conf. Series: Journal of Physics: Conf. Series **883** (2017) 012001 doi:10.1088/1742-6596/883/1/012001

We now note the approximate equivalence between the classical electron radius and the proton Compton wavelength,

$$\begin{aligned} \lambda_{\rm p} &= \frac{h}{m_{\rm p}c^2} = 1.321 \times 10^{-13}{\rm cm} \\ \simeq \quad r_{\rm e} &= \frac{e^2}{m_{\rm e}c^2} = 2.818 \times 10^{-13}{\rm cm}\,, \end{aligned} \tag{8}$$

or $\lambda_{\rm p} \simeq r_{\rm e} = \alpha_{\rm e} k_{\rm e}^{-1}$, where $k_{\rm e}^{-1} = 2\pi/\lambda_{\rm e} = \hbar/m_{\rm e}c$ is the *reduced* Compton wavelength of the electron and $\alpha_{\rm e} = e^2/\hbar c = 7.297 \times 10^{-3}$ is the fine structure constant. Performing the replacement $\lambda_{\rm p} \to r_{\rm e}$ in Eq. (3) then yields

$$\Lambda \simeq \frac{m_{\rm e}^6 G^2}{\alpha_{\rm e}^6 \hbar^4} \simeq 1.366 \times 10^{-56}{\rm cm}^{-2}\,, \tag{9}$$

or, equivalently,

$$r_{\rm e} = \frac{e^2}{m_{\rm e}c^2} \simeq (l_{\rm Pl}^2 l_{\rm dS})^{1/3}\,, \tag{10}$$

where $l_{\rm Pl} = \sqrt{\hbar G/c^3}$ denotes the *reduced* Planck length.

Remarkably, since the publication of Zel'dovich's seminal paper [7], the relations (9)-(10) have been obtained in the literature using at least *four* independent, yet not inconsistent, methods. In the present work, we review these four main, existing approaches, and briefly discuss directions for future research. The outline of this paper is as follows. In Sec. 2, the four approaches are reviewed in chronological order (i.e., the order in which they were proposed in the literature), in subsections 2.1-2.4, respectively. In Sec. 3, the implications of the relations (9)-(10) for contemporary issues in theoretical physics, including holography [10, 11] and the present day accelerated expansion of the Universe [8, 9], are considered, along with potentially novel implications for the early Universe. A brief summary of our conclusions and main results is given in Sec. 4.

2. Beyond numerology – physical arguments for the LNH

2.1. Renormalization group flow of the vacuum energy

The relations (9)-(10) were first obtained in 1993 by Nottale [12] who argued that, like other fundamental 'constants', the cosmological constant is an explicitly scale-dependent quantity, obeying a renormalization group equation. As such, its present day value may be split into a 'bare' gravitational part and a scale-dependent, quantum mechanical vacuum energy part, i.e. $\Lambda(r) = \Lambda_{\rm G} + \Lambda_{\rm Q}(r)$.

Following Zel'dovich [7], who also noted that the bare zero-point energy is unobservable, and who suggested that the *observable* contribution to the vacuum energy is given by the gravitational energy of virtual particle-antiparticle pairs, continually created and annihilated in the vacuum state,

$$E_{\rm grav} \simeq \frac{Gm^2(r)}{r}\,, \tag{11}$$

where $m(r) \simeq \hbar/(cr)$ is the effective mass of the particles at scale r, Nottale obtained the scale-dependent formula for the vacuum energy density as

$$\rho_{\rm vac}(r) \simeq \rho_{\rm Pl}\left(\frac{l_{\rm Pl}}{r}\right)^6 . \tag{12}$$

IOP Conf. Series: Journal of Physics: Conf. Series **883** (2017) 012001 doi:10.1088/1742-6596/883/1/012001

Here, $\rho_{\rm Pl} = (3/4\pi)m_{\rm Pl}/l_{\rm Pl}^3$ is the Planck density, and $m_{\rm Pl} = \sqrt{\hbar c/G}$ denotes the reduced Planck mass.

Further assuming a renormalization group equation of the form

$$\frac{d\rho_{\rm vac}}{d\ln(r)} = \gamma(\rho_{\rm vac}), \tag{13}$$

where $\gamma(\rho_{\rm vac})$ is an *unknown* function, which may be expanded as $\gamma(\rho_{\rm vac}) \simeq \gamma_0 + \gamma_1\rho_{\rm vac}$ to first order, for $\rho_{\rm vac} \leq \rho_{\rm Pl}$, yields

$$\rho_{\rm vac}(r) \simeq \rho_0 \left[1 + \left(\frac{r_0}{r}\right)^{-\gamma_1}\right], \tag{14}$$

where $\rho_0 = -\gamma_1/\gamma_0$ and r_0 is an integration constant. Comparison of Eqs. (13)-(14) then gives $\gamma_1 = -6$ and $\rho_0 = \rho_{\rm Pl}$ ($\gamma_0 = 6/\rho_{\rm Pl}$).

Thus, Nottale obtained the low-energy asymptotic value of the cosmological constant, which was found to be *scale-independent*. Next, he argued that the transition between scale-dependence and scale-independence should be identified with the Thomson scattering length (the classical electron radius), given via $\sigma_{\rm T} \simeq \pi r_{\rm e}^2$, where $\sigma_{\rm T}$ is the scattering cross-section. This is equal to the e^+e^- annihilation cross-section

$$\sigma(e^+e^-) = \pi r_{\rm e}^2 \left(\frac{m_{\rm e}c^2}{E}\right), \tag{15}$$

evaluated at $E \simeq m_{\rm e}c^2$.

In other words, $r_{\rm e}$ represents the effective electron radius, which is an energy-dependent quantity $r(E)$, evaluated at its own mass scale. The cross-section for e^+e^- pair-production at this energy scale represents the main contribution to the vacuum energy at late times. Hence, by identifying $\rho_{\rm vac} \equiv \rho_\Lambda$ and $r_0 \equiv r_{\rm e}$ in Eq. (11), he obtained the relation

$$\Lambda \simeq \frac{l_{\rm Pl}^4}{r_{\rm e}^6}, \tag{16}$$

which is equivalent to Eqs. (9)-(10).

2.2. Dark energy particles and the 'Small Number Hypothesis'

In 2008, Eqs. (9)-(10) were also obtained by Boehmer and Harko [13] in 2008, who noticed that the length scale given by the right-hand side of Eq. (4) represents the largest *stable* radius of a particle of mass

$$m_{\rm dS} \equiv \frac{\hbar}{l_{\rm dS}c} \simeq 10^{-66}\,{\rm g}, \tag{17}$$

and density $\rho_{\rm min} \simeq \rho_\Lambda$. The mass scale $m_{\rm dS}$ was previously proposed by Wesson as the minimum possible mass/energy scale in the present day Universe [14], and the existence of a minimum possible density, $\rho_{\rm min} = \rho_\Lambda/2$, for stable, charge-neutral, self-gravitating compact objects in the presence of a positive cosmological constant ($\Lambda > 0$), was shown in [15]. This result was obtained directly from the generalised Buchdahl inequalities

$$\frac{2Gm}{c^2} \geq \frac{\Lambda}{6}R^3, \quad \rho = \frac{3m}{4\pi R^3} \geq \rho_{\rm min} \equiv \frac{\Lambda c^2}{16\pi G}, \tag{18}$$

IOP Conf. Series: Journal of Physics: Conf. Series **883** (2017) 012001 doi:10.1088/1742-6596/883/1/012001

where R represents the classical radius of the compact object. Substituting $m_{\rm dS}$ into Eq. (18), we obtain

$$R \leq (l_{\rm Pl}^2 l_{\rm dS})^{1/3} \simeq 10^{-13} {\rm cm}\,. \tag{19}$$

Interestingly, it may be shown that this length scale also represents the minimum radius into which the present mass of the Universe, $m_{\rm U} \simeq m'_{\rm ds} \equiv m_{\rm Pl}^2/m_{\rm dS}$, can be compressed without exceeding the Planck density [16].

Noting the numerical coincidence between this length scale and the classical electron radius, Boehmer and Harko proposed a formal equivalence between the two on the basis of a 'Small Number Hypothesis' (SNH), which directly yields Eq. (16). However, since the reciprocal of a large number is a small number, this may be considered logically equivalent to Zel'dovich's reformulation of the LNH for $\Lambda > 0$: what differs is the empirical content, according to the substitution $\lambda_{\rm p} \rightarrow r_{\rm e}$. Thus, the work presented in [13, 16] shows that, not only is the length scale given in Eq. (19) comparable to the classical electron radius, it also represents (*i*) the maximum radius of a minimum-mass, minimum-density particle and (*ii*) the minimum radius of a maximum-mass, maximum-density particle in the observable Universe.

However, for charge-neutral, quantum mechanical particles, we may identify the radius R with the Compton wavelength $\lambda_{\rm C} = \hbar/(mc)$ (from here on, we use conventional notation and terminology, referring to the Compton wavelength and reduced Compton wavelength interchangeably), yielding [16]

$$m \geq m_{\Lambda} \equiv \frac{1}{\sqrt{2}}\sqrt{m_{\rm Pl} m_{\rm dS}} \simeq 10^{-35} {\rm g}\,. \tag{20}$$

Hence, we see that $E_{\rm dS} = m_{\rm dS}c^2$ may be interpreted as the minimum possible quantum of energy – corresponding to a de Broglie wavelength of the order of the de Sitter horizon – but *cannot* be the minimum rest mass of a stable massive particle.

The mass scale m_{Λ} (20) has several interesting properties. According to the model presented in [16], it represents the minimum possible mass of a stable, charge-neutral, quantum mechanical and self-gravitating body in the presence of dark energy ($\Lambda > 0$). With this in mind, it is notable that it is consistent with current experimental bounds on the mass of the electron neutrino obtained from Planck satellite data [9]. It may *also* be interpreted as the mass of an effective dark energy particle and its associated Compton wavelength, $l_{\Lambda} = \sqrt{2}\sqrt{l_{\rm Pl} l_{\rm dS}}$, is of the order of 0.1 mm. Thus, according to this model, the dark energy density is approximately constant over large distances, but may become granular on sub-millimetre scales. In this context, it is notable that that tentative hints of periodic variation in the gravitational field strength on precisely this length scale have recently been observed [17]. It is also the *unique* mass scale for which the Compton wavelength is equal to the gravitational turn-around radius in the presence of $\Lambda > 0$ [18].

As we shall see in Sec. 2.4, $m \simeq m_{\Lambda}$ also arises naturally in the context of the the dark-energy modified uncertainty principle (DE-UP), proposed in [19, 20], *together* with the mass scale $m \simeq m_{\rm e} \simeq \alpha_{\rm e}(m_{\rm Pl}^2 m_{\rm dS})^{1/3}$. The former corresponds to an absolute minimum, while the latter is associated with the length scale (19) via the relativistic formula for the minimum stable radius of a charged particle, $R \geq Q^2/(mc^2)$, evaluated for $Q = \pm e$.

2.3. Information theory

Also in 2008, Eqs. (9)-(10) were obtained by Beck using an information theoretic approach to the cosmological constant problem [21]. He used a system of four axioms, constructed by analogy with the Kinchin axioms of information theory [22], which describe the most 'desirable' properties of an information measure, to fix the form of the cosmological constant in terms of the other fundamental constants of nature.

It may be shown that the Kinchin axioms uniquely fix the form of the Shannon entropy, which forms the mathematical, though not the microphysical, basis of statistical mechanics and thermodynamics. Thus, by constructing an analogous approach to the cosmological constant problem, Beck attempted to fix the form of Λ in terms of the other constants of nature on an axiomatic basis, without reference to an underlying microphysical theory.

By formally replacing the dependence of the information measure I on the probabilities of events p_i by the dependence of Λ on the remaining physical constants, i.e., the fundamental coupling constants α_i, masses m_i and mixing angles s_i, he argued that the requirements of Fundamentality (L1), Boundedness (L2), Simplicity (L3) and Scale-invariance (L4) uniquely fix the form of the cosmological constant according to Eqs. (9)-(10). These axioms are constructed by analogy with the four Kinchin axioms (K1-K4) of the same names.

Specifically, K1, 'fundamentality', simply states that the information measure I should depend on the fundamental quantities, the probabilities of events p_i,

$$I = I(p_i) \, , \quad \text{(K1)} \tag{21}$$

and not on any other factors. The analogous axiom L1 for the cosmological constant is

$$\Lambda = \Lambda(\{\alpha_i\}, \{m_i\}, \{s_i\}) \, , \quad \text{(L1)} \, . \tag{22}$$

K2, 'boundedness' states that there exists a lower bound for the value of I, corresponding to the uniform distribution, $p_i = 1/N$, where N is the total number of distinct events, such that

$$I(1/N, 1/N ... 1/N) \leq I(p_1, p_2 ... p_N) \, , \quad \text{(K2)} \tag{23}$$

The analogous axiom L2 states

$$0 < \Lambda \, , \quad \text{(L2)} \tag{24}$$

where $\Lambda = 0$ is explicitly excluded. K3, 'simplicity', states that the information measure should not change if the set of events is enlarged by another set with probability zero, i.e.

$$I(p_1, p_2 ... p_N) = I(p_1, p_2 ... p_N; 0) \, , \quad \text{(K3)} \, . \tag{25}$$

The analogous axiom L3 is

$$\Lambda(\{\alpha_i\}, \{m_i\}, \{s_i\}) = \Lambda(\{\alpha_i\}, \{m_i\}, \{s_i\}; \{c_i\}) \, , \quad \text{(L3)} \, , \tag{26}$$

where the c_i are *not* fundamental constants of nature. The final axiom, K4 'invariance' is the most restrictive. It may be expressed as

$$I\left(\left\{p_{ij}^{I,II}\right\}\right) = I(\left\{p_i^I\right\}) + \sum_i p_i^I I\left\{(p^{II}(j|i)\right\}) \, , \tag{27}$$

where the superscripts I and II denote probabilities of events in different (not necessarily independent) subsystems and $p^{II}(j|i)$ is the conditional probability of event j in subsystem II, given an event i in subsystem I. $I(\left\{p^{II}(j|i)\right\})$ is the conditional information of subsystem II, in the joint system I, II, described by the probabilities $p_{ij}^{I,II} = p_i^I p^{II}(j|i)$. The meaning of K4 is that the information measure should be independent of the way in which the information is collected. We may either (a) collect information in subsystem I, then in subsystem II, or (b)

IOP Conf. Series: Journal of Physics: Conf. Series **883** (2017) 012001 doi:10.1088/1742-6596/883/1/012001

collect information in subsystem II, *assuming* a given event in subsystem I, before summing over all possible events in subsystem I, weighted by their respective probabilities p_i. Hence, there is a *scale transformation* in the space of possible information measures and probabilities, such that

$$I(\{\tilde{p}_i\}) = \tilde{I}(\{p_i\})\,, \quad \text{(K4)}\,, \tag{28}$$

where a tilde denotes transformed quantities. The analogous axiom L4 may therefore be expressed as

$$\Lambda(\{\tilde{\alpha}_i\}\,,\{\tilde{m}_i\}\,,\{\tilde{s}_i\}) = \tilde{\Lambda}(\{\alpha_i\}\,,\{m_i\}\,,\{s_i\})\,, \quad \text{(K4)}\,, \tag{29}$$

where a tilde also denotes an appropriate scale transformation: in this case, a literal rescaling of the fundamental constants of nature by an arbitrary numerical factor.

After formulating the axioms L1-L4, the argument presented in [21] relies on three specific assumptions, (i) gravitational scale invariance, (ii) a dimensional argument and (iii) electromagnetic scale invariance. The first and last of these are implicit in axiom L4, but Beck singled out the electromagnetic and gravitational interactions as being of greatest relevance to the large-scale dynamics of the early (before recombination) and late-time (after recombination) Universe, respectively. We now consider each of these assumptions in detail.

(i) Gravitational scale-invariance: In the Newtonian approximation, the the gravitational energy density of a distribution of point-like masses occupying a volume V is

$$\rho_G = -\frac{G}{V}\sum_{i,j}\frac{m_i m_j}{r_{ij}}\,, \tag{30}$$

where r_{ij} denotes the distance between the ith and j^{th} masses. Thus, if the gravitational constant G is rescaled such that

$$G \to \Gamma G\,, \tag{31}$$

where Γ is an arbitrary numerical constant, but the masses m_i, m_j and distances r_{ij} are kept the same, the energy density scales as

$$\rho_G \to \Gamma\rho_G\,. \tag{32}$$

Scale invariance of the ratio ρ_{vac}/ρ_G then requires

$$\rho_{\text{vac}} \to \Gamma\rho_{\text{vac}}\,. \tag{33}$$

Hence, $\rho_{\text{vac}} \propto G$ and we may set

$$\rho_{\text{vac}} \sim GX\,, \tag{34}$$

where X is an, as yet unknown, quantity.

(ii) The dimensional argument: On purely dimensional grounds, the unknown factor X must take the form $X \sim (c^4/\hbar^4)m^6$, where m is an arbitrary mass scale, composed (in some way) from fundamental mass scales, dimensionless coupling constants and mixing angles, according to axiom L1. (Note that this dimensional argument is unaffected even if the fundamental constants that are *not* themselves fundamental couplings, masses or mixing angles, i.e. $\hbar$ and c, are also rescaled by Γ.) Thus, without loss of generality, we may set

$$\rho_{\text{vac}} = A\frac{c^4}{\hbar^4}m_{\text{e}}^6\,, \tag{35}$$

where A is a numerical constant, which may depend on the remaining constants of nature $\{\{\alpha_i\}, \{m_i\}, \{s_i\}\}$ in any way. Likewise, A may be expressed arbitrarily in terms of $\alpha_{\rm e}$, such that

$$\rho_{\rm vac} = \frac{1}{8\pi}\alpha_{\rm e}^{\eta}\frac{c^4}{\hbar^4}m_{\rm e}^6 \,, \tag{36}$$

where η is an arbitrary function of $\{\{\alpha_i\}, \{m_i\}, \{s_i\}\}$.

(*iii*) Electromagnetic scale-invariance: Beck then argued that the most important process determining the large-scale evolution of the Universe before recombination was the Thomson scattering of electrons, associated with the scattering length $r_{\rm e} \sim \sqrt{\sigma_{\rm T}}$. Clearly, this line of reasoning is *similar* to Nottale's and Beck could also have argued *à la* Nottale that $r_{\rm e}$, interpreted as the effective cross-sectional radius for e^+e^- pair-production, is just as relevant to the large-scale dynamics of the late-time Universe. Finally, therefore, Beck claimed that, since simultaneous scale transformations of the form

$$\alpha_{\rm e} \to \Gamma\alpha_{\rm e} \,, \quad m_{\rm e} \to \Gamma m_{\rm e} \,, \tag{37}$$

leave the Thomson scattering cross-section $\sigma_{\rm T}$ invariant, scale-invariance of the ratio $\Lambda^{-1}/\sigma_{\rm T}$ requires that Λ be dependent on the ratio $m_{\rm e}/\alpha_{\rm e}$ *only*. This fixes the remaining free constant, $\eta = -6$, and the factor of $(8\pi)^{-1}$ in Eq. (36) is chosen so as to match the numerical factor in Eq. (6), giving the 'simplest' result:

$$\rho_\Lambda = \frac{\Lambda c^2}{8\pi G} \equiv \rho_{\rm vac} = \frac{1}{8\pi}\frac{c^4}{\hbar^4}\left(\frac{m_{\rm e}}{\alpha_{\rm e}}\right)^6 . \tag{38}$$

It is straightforward to check that this is relation is equivalent to Eqs. (9)-(10) and (16).

However, it must be noted that an obvious problem with Beck's approach is that, unlike the probabilities of events, which are *all* dimensionless, the fundamental quantities on which he bases the expression for the cosmological constant Λ are inequivalent in this respect. Though *non*-gravitational coupling constants may be expressed in dimensionless form using $\hbar$ and c, Newton's constant G cannot. Similarly, the fundamental masses cannot be expressed in dimensionless form without the use of the Planck mass, $m_{\rm Pl} \propto 1/\sqrt{G}$. A related problem concerns the fact that, if the ultimate origin of the point-like masses considered in Eq. (30) are *fundamental* masses, these should also be rescaled at the same time as G, which changes the scaling of $\rho_{\rm vac}$ with Γ, dramatically.

Overall, one can say that it does not make physical sense to rescale *dissimilar* quantities (i.e., those with inequivalent dimensions) with the same scale factor: only similar quantities should be rescaled in this way. A self-consistent set of rescalings, which leave dimensionless quantities constructed from Eq. (16) explicitly invariant (in line with axiom L4), would then be

$$\Lambda^{-1} \to \Gamma\Lambda^{-1} \,, \quad \sigma_{\rm T} \to \Gamma\sigma_{\rm T} \,, \quad l_{\rm Pl}^2 \to \Gamma l_{\rm Pl}^2 \,. \tag{39}$$

Alternatively, instead of requiring the absolute scale independence of Λ, or even the relative scale independence suggested by Eq. (39), we may instead require *holographic* relationships between the bulk and the boundary of the Universe to be preserved by our expression for Λ, written in terms of the remaining constants of nature. In the following section, we summarize the results of a new approach to deriving the expressions (9)-(10) and (16), which automatically implements the holographic principle [10, 11].

2.4. Minimum length uncertainty relations in a dark energy Universe

A further attempt to explain the connection between cosmological and atomic scales in the LNH, in terms of the stability of fundamental particles in the presence of dark energy, was presented in a series of recent papers [19, 20]. (See also [23, 24] for extensions to modified gravity theories.) This approach considered the status of minimum length uncertainty relations (MLURs), motivated by quantum gravity phenomenology (see [25, 26] for contemporary reviews), in the presence of a vacuum energy density ρ_Λ given by Eq. (6).

In [19], a new form of MLUR, dubbed the 'dark energy uncertainty principle' or DE-UP for short, which explicitly includes the de Sitter scale $l_{dS} \sim 1/\sqrt{\Lambda}$, as well as the Planck scale, was proposed. The general form of the DE-UP proposed in [19] is

$$\begin{aligned} \Delta x_{\text{total}}(\Delta v, r, m) &= \Delta x_{\text{canon.}}(\Delta v, r, m) + \Delta x_{\text{grav}}(r, m) \\ &\geq \Delta x(\Delta v) + \Delta x_{\text{recoil}}(\Delta v, r, m) \\ &+ \Delta x_{\text{grav}}(r, m) \\ &\geq (\Delta x_{\text{canon.}})_{\text{min}}(r, m) \\ &+ \Delta x_{\text{grav}}(r, m) \,. \end{aligned} \tag{40}$$

Here, $(\Delta x_{\text{canon.}})_{\text{min}}$ denotes the minimum possible canonical quantum uncertainty of a wave packet that has been freely evolving for a time $t = r/c$. By explicitly solving the free-particle Schrödinger equation in the Heisenberg picture, it is straightforward to show that, for a particle of mass m, the minimum canonical uncertainty is

$$(\Delta x_{\text{canon.}})_{\text{min}} \simeq \sqrt{\lambda_C r} \,, \tag{41}$$

where $r = ct$ and λ_C is the Compton wavelength [27, 28]. This expression can also be obtained by considering a gedanken experiment, originally due to Salecker and Wigner, in which a massive particle is used to 'measure' a distance $r \simeq ct$ by means of the emission and reabsorption of a photon [29]. By minimising the sum of the first two terms on the top line of Eq. (40), where Δx represents the canonical Heisenberg uncertainty and $\Delta x_{\text{recoil}} \simeq \Delta vt \equiv \Delta pr/(mc)$ is the additional uncertainty due to recoil, with respect to either Δv or m, we obtain Eq. (41) directly.

The term Δx_{grav} represents an additional contribution to the total uncertainty, due to the superposition of gravitational field states which are (in turn) induced by the superposition of position states associated with m. From a relativistic perspective, this may be considered *equivalent* to the uncertainty associated with the superposition of space-time geometries in the quantum gravity regime. In [19], it was conjectured that this superposition is influenced by two factors: (i) the mass of the particle, and (ii) the presence of the dark energy density, or, equivalently, of a finite horizon for the wave function centre of mass, $l_{dS} \sim 1/\sqrt{\Lambda}$. Taking both these factors into account, the conjectured form of Δx_{grav} proposed in [19] was

$$\Delta x_{\text{grav}} \simeq \frac{1}{\sqrt{\Lambda}} \frac{Gm}{c^2 r} \simeq \frac{l_{\text{Pl}}^2 l_{\text{dS}}}{\lambda_C r} \,. \tag{42}$$

This was combined with the limit on the mass/radius ratio for stable, charged, compact objects originally obtained by Bekenstein [30], and later generalised by Boehmer and Harko for $\Lambda > 0$ [31], i.e.

$$R \geq \frac{Q^2}{mc^2} + \text{h.o.t.}(G, \Lambda \,. \,. \,) \,. \tag{43}$$

To leading order, Eq. (43) simply recovers the well-known formula for the classical radius of a charged particle obtained from special relativity, but this remains rigorously valid in the weak-field limit of general relativity, even in the presence of dark energy [30, 31]. By identifying the *total* minimum quantum mechanical uncertainty $(\Delta x_{\rm total})_{\rm min}$, including both canonical quantum and gravitational/dark energy-induced terms, with the minimum radius R obtained from the Bekenstein bound, the following inequality is obtained

$$\frac{Q^2}{mc^2} \leq (l_{\rm Pl}^2 l_{\rm dS})^{1/3}\,. \tag{44}$$

Thus, according to the model presented in [19], Eq. (44) gives the maximum possible charge-squared to mass ratio for a stable, charged, self-gravitating and quantum mechanical object. Assuming saturation of this bound is equivalent to assuming the existence of a particle in nature that simultaneously saturates both the classical (general-relativistic) *and* quantum mechanical (MLUR) stability constraints. Remarkably, comparison with Eq. (10) reveals that saturation is obtained for the electron charge-squared to mass ratio. Equivalently, we see that evaluating (44) for $Q = \pm e$, yields

$$\begin{aligned} m &\geq \alpha_e (m_{\rm Pl}^2 m_{\rm dS})^{1/3} = 7.332 \times 10^{-28}\,{\rm g} \\ &\simeq \; m_e = 9.109 \times 10^{-28}\,{\rm g}\,. \end{aligned} \tag{45}$$

Hence, if the electron were more highly charged (with the same mass $m_{\rm e}$) or any less massive (with the same charge e), a combination of electrostatic and dark energy repulsion would destabilize its Compton wavelength [19].

In [20], a full *physical* derivation of the dark energy-modified uncertainty principle (DE-UP), proposed in [19] is given, in which it is shown that the de Sitter length-dependent term arises as a direct consequence of the existence of a finite horizon $r_{\rm H}(\tau_0) \simeq l_{\rm dS}$, where τ_0 is the present age of the Universe. In addition, it is shown that the DE-UP defined by Eqs. (40) and (41)-(42) naturally incorporates the mass bound for neutral particles, Eq. (20), in addition to that for charged particles, Eq. (45). In particular, it is straightforward to show that requiring every potentially observable length-scale, i.e. $(\Delta x_{\rm canon.})_{\rm min}$, $\Delta x_{\rm grav}$ and the 'probe' distance r, to be super-Planckian, automatically implies the existence of a minimum mass in nature, $m_\Lambda \simeq \sqrt{m_{\rm Pl} m_{\rm dS}}$. Alternatively, beginning with the bound (20), obtained by combining both general-relativistic and quantum mechanical effects, we obtain $l_{\rm Pl}$ as a limiting resolution for physical measurements of length within the DE-UP framework.

Finally, we note that, using Salam's theory of strong gravity [32, 33, 34, 35] as an *effective* theory for modeling quark confinement, analogous arguments where applied to charged, strongly interacting particles, in which the energy density associated with the 'strong de Sitter radius' was identified with the 'bag constant' of the MIT bag model for confined nuclear matter, $B \simeq 10^{14}$ gcm^{-1} [24]. These arguments successfully predicted the correct order of magnitude value for the charge-squared to mass ratio of the up quark, the lightest known strongly interacting and quantum mechanical particle, with charge $Q = \pm 2e/3$ [24].

3. Implications of a connection between "dark" and "light" physics

3.1. Holography

It is straightforward to see that, for any particle that minimizes the total uncertainty given by the DE-UP, Eqs. (40) and (41)-(42), a holographic relation holds between the bulk and the boundary of the Universe, namely

$$\left(\frac{(\Delta x_{\rm total})_{\rm min}}{l_{\rm dS}}\right)^3 = \frac{l_{\rm Pl}^2}{l_{\rm dS}^2} = N = 1.030 \times 10^{122}\,. \tag{46}$$

Hence, the number of Planck sized 'bits' on the de Sitter boundary is equal to the number of minimum-volume 'cells', $V_{\rm cell} \sim (\Delta x_{\rm total})_{\rm min}^3$, in the bulk [19]. As mentioned previously, in Sec.

2.2, $R \simeq (\Delta x_{\text{total}})_{\min}$ may also be interpreted as the *classical* radius of a 'particle' with both minimum energy, $E_{\text{dS}} = m_{\text{dS}}c^2$, and minimum density, $\rho_{\min} = \rho_\Lambda/2$, i.e.

$$\rho \simeq \frac{m_{\text{dS}}}{R^3} \geq \rho_{\min} \simeq \frac{\Lambda c^2}{G} \iff R \leq (\Delta x_{\text{total}})_{\min} \simeq (l_{\text{Pl}}^2 l_{\text{dS}})^{1/3} . \tag{47}$$

Although a massive particle with rest energy E_{dS} would be unstable due to the repulsive effects of dark energy, this *may* correspond to the energy of a photon with maximum wavelength, $\lambda \simeq l_{\text{dS}}$. Thus, $(\Delta x_{\text{total}})_{\min}$ may also be interpreted as the classical radius of a localized, minimum-energy photon. A space-filling 'sea' of such photons would have the same energy density as the dark energy field [16].

In addition, we may consider a maximum-mass, maximum-density state, for which $\rho \simeq \rho_{\text{Pl}}$ and the total energy is $E'_{\text{dS}} \simeq m'_{\text{dS}}c^2$ ($m'_{\text{dS}} \equiv m_{\text{Pl}}^2/m_{\text{dS}}$). The classical radius thus obtained corresponds to the smallest possible volume within which the total mass of the present day horizon may be confined, without exceeding the Planck density. We then have [19]

$$\rho \simeq \frac{m'_{\text{dS}}}{R^3} \leq \rho_{\text{Pl}} \simeq \frac{c^5}{\hbar G} \iff R \geq (\Delta x_{\text{total}})_{\min} \simeq (l_{\text{Pl}}^2 l_{\text{dS}})^{1/3} . \tag{48}$$

The length scale $(\Delta x_{\text{total}})_{\min} \simeq (l_{\text{Pl}}^2 l_{\text{dS}})^{1/3}$ therefore corresponds to at least three interesting physical scenarios. It may be interpreted as *(i)* the classical radius of a 'particle' with both minimum energy and minimum energy density, *(ii)* the classical radius of a 'particle' with both maximum energy and maximum energy density, and *(iii)* the classical radius/minimum total uncertainty of the electron, which saturates the charged particle stability bound (44). In the context of the DE-UP model [19, 20], *all three* interpretations satisfy the general holographic relation, Eq. (46).

3.2. Universal expansion

As pointed out in [16], the minimum mass $m_\Lambda \simeq \sqrt{m_{\text{Pl}} m_{\text{dS}}}$ (20) may also be interpreted as the mass of an *effective* dark energy particle. Since, in this model, even random quantum fluctuations reduce the inter-particle distance between nearest neighbours to less than $\lambda_C(m_\Lambda) = l_\Lambda \simeq \sqrt{l_{\text{Pl}} l_{\text{dS}}}$, it follows that the pair-production of "dark" minimum-mass particles is capable of driving the present day accelerated expansion of the Universe. In short, if space is 'full' of dark energy particles, with mean inter-particle distance $l_\Lambda \sim 0.1$ mm, the pair-production necessitated by quantum mechanics *requires* a concomitant expansion of space [16, 24].

Let us assume that the probability of a single (holographic) spatial cell 'pair-producing' within a time interval $\Delta\tau = t_{\text{Pl}} = l_{\text{Pl}}/c$, due to the pair-production of dark energy particles, is given by

$$P(\Delta V = +V_{\text{cell}} | V_0 = V_{\text{cell}}, \Delta\tau = t_{\text{Pl}}) = N^{-1/2}$$
$$= \frac{V_{\text{Pl}}}{V_{\text{cell}}} = \frac{l_{\text{Pl}}}{l_{\text{dS}}} \simeq \left(\frac{\hbar G \Lambda}{3c^3}\right)^{1/2} \simeq 9.851 \times 10^{-62} , \tag{49}$$

where V_0 denotes the initial volume at the initial time. This leads naturally to a de Sitter-type expansion, modelled by the differential equation

$$\frac{da^3}{d\tau} \simeq \frac{N^{-1/2} a^3}{t_{\text{Pl}}} = \frac{l_{\text{Pl}}}{l_{\text{dS}}} \frac{a^3}{t_{\text{Pl}}} , \tag{50}$$

or, equivalently,

$$\frac{da}{d\tau} \simeq c\sqrt{\frac{\Lambda}{3}} a , \quad a(\tau) \simeq a_0 e^{-c\sqrt{\Lambda/3}\tau} . \tag{51}$$

Since the production of a single dark energy energy particle requires the production of $n_{\rm cell} = V_\Lambda/V_{\rm cell} \simeq l_\Lambda^3/(l_{\rm Pl}^2 l_{\rm ds}) = N^{1/4}$ cells of space, this implies that the probability of a dark energy particle pair-producing within a single Planck time is given by

$$\begin{aligned} P(\Delta V &= +V_\Lambda | V_0 = V_\Lambda, \Delta t = t_{\rm Pl}) \simeq N^{-3/4} \\ &= \left(\frac{l_{\rm ds}}{l_{\rm Pl}}\right)^{-3/2} \simeq 10^{-91} \,. \end{aligned} \quad (52)$$

However, since there are $n_{\rm DE} \simeq l_{\rm dS}^3/l_\Lambda^3 = N^{3/4}$ dark energy particles within the de Sitter horizon, this means that one dark energy particle is produced, *somewhere* in the observable Universe, during every Planck time interval. This rate of pair-production is capable of giving rise to the accelerated Universal expansion observed at the present epoch.

In this model, the observed vacuum energy is really the energy associated with the dark energy field, for which $\lambda_{\rm C}(m_\Lambda) = l_\Lambda$ provides a natural a UV cut-off for the field modes, yielding

$$\rho_{\rm vac} \simeq \frac{\hbar}{c}\int_{1/l_{\rm dS}}^{1/l_\Lambda} \sqrt{k^2 + \left(\frac{2\pi}{l_\Lambda}\right)^2} {\rm d}^3k \simeq \frac{m_{\rm Pl} l_{\rm Pl}}{l_\Lambda^4} \simeq \frac{\Lambda c^2}{G} \simeq 10^{30} \ {\rm gcm}^{-3} \,. \quad (53)$$

The field itself remains 'trapped' in a Hagedorn-type phase, in which any increase in kinetic energy, *even that caused by random collisions between neighbouring dark energy particles due to quantum uncertainty*, results in pair-production, rather than an increase in temperature.

The temperature associated with the field therefore remains constant, on large scales, and is comparable to the present day temperature of the CMB,

$$T_\Lambda \equiv \frac{m_\Lambda c^2}{8\pi k_{\rm B}} \simeq 2.27 \ {\rm K} \simeq T_{\rm CMB} = 2.73 \ {\rm K} \,. \quad (54)$$

Here the factor of 8π is included by analogy with the expression for the Hawking temperature,

$$T_{\rm H} \equiv \frac{c^2}{8\pi k_{\rm B}} \frac{m_{\rm Pl}^2}{m} \,, \quad (55)$$

so that $T_\Lambda \equiv T(m_\Lambda) = T_{\rm H}(m'_\Lambda)$, where $m'_\Lambda \equiv m_{\rm Pl}^2/m_\Lambda$ is the dual mass. Though this may seem like another 'miraculous' coincidence, *à la* Dirac, in the dark energy model implied by the DE-UP it is simply a re-statement of the standard 'coincidence problem' of cosmology, i.e. the Universe begins a phase of accelerated expansion when $r_{\rm U} \simeq l_{\rm dS}$, at which point $\Omega_{\rm M} \simeq \Omega_\Lambda$ and, hence, $T_{\rm CMB} \simeq T_\Lambda$. The question remains, why do we live at precisely this epoch? However, no *new* coincidences are required, in order to 'explain' Eq. (54).

Again, we may apply analogous arguments to strongly interacting particles by using Salam's theory of strong gravity as an *effective* theory for confined (and deconfined) nuclear matter. These predict a genuine Hagedorn temperature for the quark-gluon plasma of order $T_{\rm Hag} \simeq N^{1/10} T_\Lambda \simeq 10^{12}$ K [24].

3.3. Additional cosmological implications

The holographic relation Eq. (46) remains valid for all epochs, prior to the present day, under the substitution $l_{\rm dS} \to r_{\rm H}(\tau)$, where $r_{\rm H}(\tau)$ is the physical horizon radius at cosmic time τ. However, under these circumstances, we note that the DE-UP model naturally implies both a time-dependent minimum mass for neutral particles *and* a time-dependent maximum charge-squared to mass ratio for charged particles, i.e.

$$m_\nu(\tau) \geq \sqrt{m_{\rm Pl} m_{\rm H}(\tau)} \,, \quad (56)$$

where $m_{\rm H}(\tau) \equiv \hbar/(r_{\rm H}(\tau)c)$, and

$$\frac{e^2}{m_e}(\tau) \leq c^2 (l_{\rm Pl}^2 r_{\rm H}(\tau))^{1/3}, \tag{57}$$

respectively [20]. Equation (57) corresponds to a three-dimensional, time-dependent, holographic cell radius

$$(\Delta x_{\rm total})_{\rm min}(\tau) \simeq (l_{\rm Pl}^2 r_{\rm H}(\tau))^{1/3}. \tag{58}$$

This is similar to the MLUR for an expanding Universe suggested by Ng [36], but with the cosmological horizon $r_{\rm H}(\tau)$ in place of the Hubble horizon $\mathcal{H}(\tau)/c$.

Though highly speculative, these relations imply an interesting form of 'unification' in the early Universe, with all masses tending to the Planck mass and all charges tending to the Planck charge, $q_{\rm Pl} = \sqrt{\hbar c}$ as $r_{\rm H}(\tau) \to l_{\rm Pl}$. Alternatively, if the Planck-density threshold limits the radius of the 'initial' big bang horizon such that $r_{\rm H}(\tau) \geq (l_{\rm Pl}^2 l_{\rm dS})^{1/3}$, as suggested by the results obtained in [19], the minimum holographic cell radius will be of order

$$R_{\rm min} \simeq (l_{\rm Pl}^8 l_{\rm dS})^{1/9} \simeq 10^{-26} \text{ cm}. \tag{59}$$

Finally, we note that the length and time scales associated with the mass $m_\nu(\tau)$ (56), i.e.

$$r_\nu(\tau) = ct_\nu(\tau) \equiv \frac{\hbar}{m_\nu(\tau)c}, \tag{60}$$

also satisfy the 'four-dimensional' holographic relation

$$\left(\frac{r_{\rm H}(\tau)}{r_\nu(\tau)}\right)^4 = \frac{r_{\rm H}^2(\tau)}{l_{\rm Pl}^2} = N(\tau) \leq 1.030 \times 10^{122}. \tag{61}$$

Hence, the DE-UP model strongly suggests time-variation of either, or both, e and $m_{\rm e}$, assuming that $\{G, c, \hbar, \Lambda\}$ are genuine universal constants. Similar arguments apply to the mass of the lightest neutral particle, previously identified with the mass of the electron neutrino, m_ν.

For models involving temporal and/or spatial variation of fundamental constants, the situation is even more complicated, and it may be extremely difficult, in practice, to distinguish variation in e and/or $m_{\rm e}$, and m_ν, from other effects. Particular classes of models in which the variation of physical constants should, automatically, imply a modification of the DE-UP formulae, Eqs. (40) and (41)-(42), include those with a running gravitational coupling [37, 38, 39, 40], variable speed of light [41, 42, 43, 44, 45, 46, 47, 48], or dynamical dark energy field [49, 50, 51, 52]. (See also [53, 54, 55, 56, 57] for current bounds on varying α_e theories, including their effects on cosmic string phenomenology [58, 59] and [60, 61, 62, 63, 64] for more general models involving variations of multiple physical constants.)

In fact, several models incorporating non-minimal couplings between dark energy and the electromagnetic sector have already been proposed in the literature, as solutions to problems in contemporary cosmology [65, 66, 67, 68, 69]. Although a thorough analysis of the cosmological implications of the DE-UP has not yet been attempted, the cosmological implications of $\Lambda \propto \alpha_{\rm e}^{-6}$ cosmology were investigated in [70, 71], in the context of a time-varying fine structure constant. Nonetheless, its unusual predictions suggest that future observations and/or analysis of currently available data may be capable of falsifying the model [20].

4. Discussion

We have shown that the relations (9)-(10) and (16), which are equivalent to Zel'dovich's reformulation of Dirac's Large Number Hypothesis for a Universe with $\Lambda > 0$, under the identification $m \simeq m_e/\alpha_e$, are well motivated from a number of theoretical perspectives. Each of these goes beyond numerology and aims to base the seemingly incredible coincidences noted by Dirac, Zel'dovich and others on firm *physical* arguments. Specifically, we have outlined four sets of independent, yet not necessarily incompatible arguments, given in the literature, which give rise to the relation $\Lambda \simeq l_{\rm Pl}^4/r_e^6$ (16), where $r_e = e^2/(m_e c^2)$ is the classical electron radius.

The first, proposed by Nottale [12] (1993), is based on the assumption that the vacuum energy density is dominated by the gravitational energy associated with e^+e^- pair production. By perturbatively expanding the renormalisation group equation for $\rho_{\rm vac}$, and identifying the transition from running to scale-independence with the energy scale $E \simeq m_e c^2$, he obtained Eq. (16) directly.

The second, proposed by Boehmer and Harko [13] (2008), identifies the maximum stable radius of a minimum-mass, minimum-density 'particle', in the presence of a positive cosmological constant $\Lambda > 0$, with the classical electron radius via a 'Small Number Hypothesis'. Since the reciprocal of a large number is a small number, this is logically equivalent to Zel'dovich's reformulation of the LNH with $m \simeq m_e/\alpha_e$. However, the important *physical* content of this work is the realisation that the cosmological constant automatically implies the existence of a minimum density for stable compact objects in nature. This result arises rigorously from the generalized Buchdahl inequalities [15] and, following a similar analysis, the minimum charge-squared to mass ratio for electrically charged particles can also be obtained. Identifying this with the alternative expression for r_e, obtained from the SNH, *also* yields (16).

The third method, proposed by Beck [21] (2008) follows an axiomatic approach, based on analogy with the Kinchin axioms of information theory [22]. These uniquely fix the form of the Shannon entropy, which forms the mathematical basis of statistical mechanics and thermodynamics, *without* making any assumptions about the underlying microphysical basis of these theories. Likewise, Beck's approach aims to uniquely fix the form of Λ without reference to an underlying microphysical model. By formally replacing the dependence of the Shannon information measure I on the probabilities of events p_i by the dependence of Λ on the fundamental constants of nature (i.e., the fundamental coupling constants, masses and mixing angles), Eq. (16) was obtained from the requirements of 'Fundamentality', 'Boundedness', 'Simplicity' and 'Scale-invariance' [21].

A fourth derivation of Eq. (16) has been proposed in a recent series of papers by Burikham, Cheamsawat, Harko, Lake and Paterek [19, 20, 23, 24] (2016-2017). This is based on the construction of a dark energy-modified minimum length uncertainty relation, dubbed the dark energy uncertainty principle, or DE-UP for short, which may be combined with *classical* minimum mass, radius and/or density bounds to yield stability conditions for self-gravitating, compact and quantum mechanical objects. Applying the DE-UP to charge-neutral particles recovers the minimum-mass bound previously obtained in [16], which is consistent with current bounds on the mass of the electron neutrino [9], whereas applying it to charged particles with $Q = \pm e$ yields the electron mass, in accordance with Eq. (16).

The form of the DE-UP proposed in [19] leads naturally to holographic relation between the bulk and the de Sitter horizon, in which the number of minimum-volume 'cells' equals the number of Planck-sized 'bits' on the boundary [72, 73, 74]. However, it must be noted that, in order to maintain this relation for $\tau \ll \tau_0$, where $\tau_0 \simeq t_{\rm dS} = l_{\rm dS}/c$, we must substitute $l_{\rm dS} \to r_{\rm H}(\tau)$, where $r_{\rm H}(\tau)$ is the physical horizon radius at cosmic time τ. Performing the same substitution in the DE-UP – the validity of which is also supported by the physical arguments proposed in [20] – therefore gives rise to time-variation of the minimum mass for charge-neutral particles *and* of the maximum charge-squared to mass ratio for charged particles.

At present, further analysis is required to determine whether the predictions of the DE-UP, which gives rise to a natural implementation of the LNH at the present epoch, as well as to a natural description of late-time accelerated expansion in terms of dark energy particles (see Sec. 3.2), is compatible with existing cosmological data. However, it must be noted that the present model implicitly assumes that G, c, $\hbar$ and Λ are genuine universal constants which do not vary in time. Running of the gravitational coupling, as recently claimed in [37, 38, 39, 40], or additional time-dependence in any of the parameters $\{G, c, \bar{h}\Lambda\}$, may dramatically alter these predictions. With this in mind, we note that a model in which $\Lambda \propto \alpha_e^{-6}$, in the context of varying α_e cosmology, was considered in [70, 71], whereas alternative ways of incorporating the effects of Universal expansion/dark energy on the uncertainty principle were considered in [75, 76].

Finally, we may consider the implications of a connection between "dark" and "light" physics, suggested by Eq. (16), for the physics of black holes. *A priori*, Eqs. (9)-(10) and (16) say nothing about black holes, yet if, as claimed in [19, 20], the ultimate origin of these relations is the DE-UP, it is by no means clear whether this even applies to objects with masses $m \geq m_{\rm Pl}$. In general, the form of positional uncertainty (if any) obeyed by the centre-of-mass of a black hole, remains an outstanding problem in contemporary theoretical physics. (See for [77, 78, 79, 80, 81, 82] for recent works in this direction.)

Nonetheless, it is certainly worthwhile to attempt to extend the DE-UP into this region, which may be done naïvely by simply replacing the rest mass m with the 'dual' ADM mass $m \to m'_{\rm ADM} \equiv m_{\rm Pl}^2/m_{\rm ADM} \simeq m_{\rm Pl}^2/(m + m_{\rm Pl}^2/m)$. This gives rise to a unified Compton-Schwarzschild line connecting the black hole and particle regimes (see [77, 78, 79, 80, 81, 82] and [83, 84]). Since the DE-UP naturally implements holography in the $m \leq m_{\rm Pl}$ regime, it may be hoped that the extended version maintains it for $m \geq m_{\rm Pl}$, which may have profound implications for the black hole information loss paradox [85, 86, 87]. In this context, reassessing Beck's information-theoretic approach, subject to holographic constraints, may prove particularly fruitful.

Acknowledgments

M.L. thanks the Institute for Fundamental Study (IF), Phitsanulok, Thailand, the Yukawa Institute for Theoretical Physics (YITP), Kyoto, Japan, and the Asia-Pacific Center for Theoretical Physics (APCTP), Pohang, Korea, for their generous support of the 6th IF+YITP International Symposium, held at Naresuan University in August 2016. He is supported by a Naresuan University Research Fund individual research grant.

References

[1] P. A. M. Dirac, *The cosmological constants*, Nature 139 (1937): 323.
[2] P. A. M. Dirac, *A new basis for cosmology*, Proc. R. Soc. Lond. A Vol. 165. No. 921 (1938).
[3] P. A. M. Dirac, *Cosmological Models and the Large Numbers Hypothesis* Proc. R. Soc. Lond. A 1974 338 439-446, DOI: 10.1098/rspa.1974.0095 (1974).
[4] P. A. M. Dirac, *The Large Numbers Hypothesis and the Einstein Theory of Gravitation* Proc. R. Soc. Lond. A 1979 365 19-30, DOI: 10.1098/rspa.1979.0003 (1979).
[5] S. Ray, U. Mukhopadhyay and P. Pratim Ghosh, *Large Number Hypothesis: A Review*, [arXiv:0705.1836 [gr-qc]].
[6] J. P. Ostriker and P. J. Steinhardt, *Cosmic concordance*, astro-ph/9505066.
[7] Y. B. Zel'dovich, A. Krasinski and Y. B. Zeldovich, *The Cosmological constant and the theory of elementary particles*, Sov. Phys. Usp. **11**, 381 (1968) [Gen. Rel. Grav. **40**, 1557 (2008)] [Usp. Fiz. Nauk **95**, 209 (1968)]. doi:10.1007/s10714-008-0624-6, 10.1070/PU1968v011n03ABEH003927
[8] M. Betoule *et al.* [SDSS Collaboration], *Improved cosmological constraints from a joint analysis of the SDSS-II and SNLS supernova samples*, Astron. Astrophys. **568**, A22 (2014) doi:10.1051/0004-6361/201423413 [arXiv:1401.4064 [astro-ph.CO]].
[9] P. A. R. Ade *et al.* [Planck Collaboration], *Planck 2013 results. I. Overview of products and scientific results*, arXiv:1303.5062 [astro-ph.CO]; P. A. R. Ade *et al.* [Planck Collaboration], *Planck 2013 results. XVI. Cosmological parameters*, arXiv:1303.5076 [astro-ph.CO]. P. A. R. Ade *et al.* [Planck Collaboration], *Planck*

intermediate results. XXII. Frequency dependence of thermal emission from Galactic dust in intensity and polarization arXiv:1405.0874 [astro-ph.GA]; P. A. R. Ade *et al.* [Planck Collaboration], *Planck intermediate results. XXIX. All-sky dust modelling with Planck, IRAS, and WISE observations* arXiv:1409.2495 [astro-ph.GA].

[10] G. 't Hooft, *The Holographic principle: Opening lecture*, Subnucl. Ser. **37**, 72 (2001) doi:10.1142/9789812811585 0005 [hep-th/0003004].

[11] R. Bousso, *The Holographic principle*, Rev. Mod. Phys. **74**, 825 (2002) doi:10.1103/RevModPhys.74.825 [hep-th/0203101].

[12] L. Nottale, *Mach's Principle, Dirac's Large Number Hypothesis and the Cosmological Constant Problem* (preprint), https://www.luth.obspm.fr/ luthier/nottale/arlambda.pdf (1993).

[13] C. G. Boehmer and T. Harko, *Physics of dark energy particles*, Found. Phys. **38**, 216 (2008) doi:10.1007/s10701-007-9199-4 [gr-qc/0602081].

[14] P. S. Wesson, *Is mass quantized?*, Mod. Phys. Lett. A **19**, 1995 (2004).

[15] C. G. Boehmer and T. Harko, *Does the cosmological constant imply the existence of a minimum mass?*, Phys. Lett. B **630**, 73 (2005) doi:10.1016/j.physletb.2005.09.071 [gr-qc/0509110].

[16] P. Burikham, K. Cheamsawat, T. Harko and M. J. Lake, *The minimum mass of a spherically symmetric object in D-dimensions, and its implications for the mass hierarchy problem*, Eur. Phys. J. C **75**, no. 9, 442 (2015) [arXiv:1508.03832 [gr-qc]].

[17] L. Perivolaropoulos, *Sub-millimeter Spatial Oscillations of Newton's Constant: Theoretical Models and Laboratory Tests*, arXiv:1611.07293 [gr-qc].

[18] S. Bhattacharya, K. F. Dialektopoulos, A. E. Romano, C. Skordis and T. N. Tomaras, *The maximum sizes of large scale structures in alternative theories of gravity*, arXiv:1611.05055 [astro-ph.CO].

[19] P. Burikham, K. Cheamsawat, T. Harko and M. J. Lake, *The minimum mass of a charged spherically symmetric object in D dimensions, its implications for fundamental particles, and holography*, Eur. Phys. J. C **76**, no. 3, 106 (2016) [arXiv:1512.07413 [gr-qc]].

[20] M. J. Lake and T. Paterek, *Minimum length uncertainty relations for a dark energy Universe* (in preparation).

[21] C. Beck, *Axiomatic approach to the cosmological constant*, Physica A **388**, 3384 (2009).

[22] A. I. Khinchin, *Mathematical Foundations of Information Theory*, Dover Publ., New York (1957).

[23] P. Burikham, T. Harko and M. J. Lake, *Mass bounds for compact spherically symmetric objects in generalized gravity theories*, Phys. Rev. D **94**, no. 6, 064070 (2016) doi:10.1103/PhysRevD.94.064070 [arXiv:1606.05515 [gr-qc]].

[24] P. Burikham, T. Harko and M. J. Lake, *The QCD mass gap and quark deconfinement scales as mass bounds in strong gravity*, arXiv:1705.11174 [hep-th].

[25] S. Hossenfelder, *Minimal Length Scale Scenarios for Quantum Gravity*, Living Rev. Rel. **16**, 2 (2013).

[26] L. J. Garay, *Quantum gravity and minimum length*, Int. J. Mod. Phys. A **10**, 145 (1995) [gr-qc/9403008].

[27] X. Calmet, M. Graesser and S. D. H. Hsu, *Minimum length from quantum mechanics and general relativity*, Phys. Rev. Lett. **93**, 211101 (2004).

[28] X. Calmet, M. Graesser and S. D. H. Hsu, *Minimum length from first principles*, Int. J. Mod. Phys. D **14**, 2195 (2005).

[29] H. Salecker and E. P. Wigner, *Quantum limitations of the measurement of space-time distances*, Phys. Rev. **109**, 571 (1958). doi:10.1103/PhysRev.109.571

[30] J. D. Bekenstein, *Hydrostatic Equilibrium and Gravitational Collapse of Relativistic Charged Fluid Balls*, Phys. Rev. D **4**, 2185 (1971).

[31] C. G. Boehmer and T. Harko, *Minimum mass-radius ratio for charged gravitational objects*, Gen. Rel. Grav. **39**, 757 (2007).

[32] C. J. Isham, A. Salam and J. A. Strathdee, *F-dominance of gravity*, Phys. Rev. D **3**, 867 (1971). doi:10.1103/PhysRevD.3.867

[33] C. J. Isham, A. Salam and J. A. Strathdee, *Infinity suppression gravity modified quantum electrodynamics*, Phys. Rev. D **3**, 1805 (1971). doi:10.1103/PhysRevD.3.1805

[34] C. J. Isham, A. Salam and J. A. Strathdee, *2+ nonet as gauge particles for sl(6,c) symmetry*, Phys. Rev. D **8**, 2600 (1973). doi:10.1103/PhysRevD.8.2600

[35] C. J. Isham and M. C. Tucker, *A* $U(3)$ *Version of the f-g Theory*, Phys. Rev. D **10**, 3219 (1974). doi:10.1103/PhysRevD.10.3219

[36] Y. J. Ng, *Quantum Foam, Gravitational Thermodynamics, and the Dark Sector*, J. Phys. Conf. Ser. **845**, no. 1, 012001 (2017) doi:10.1088/1742-6596/845/1/012001 [arXiv:1701.00017 [gr-qc]].

[37] A. Gómez-Valent, J. Solà and J. d. C. Pérez, *Running vacuum versus the* ΛCDM, arXiv:1605.06448 [gr-qc].

[38] H. Fritzsch, R. C. Nunes and J. Solà, *Running vacuum in the Universe and the time variation of the fundamental constants of Nature*, arXiv:1605.06104 [hep-ph].

[39] J. Solà, A. Gómez-Valent and J. d. C. Pérez, *First evidence of running cosmic vacuum: challenging the*

concordance model, arXiv:1602.02103 [astro-ph.CO].
[40] J. Solà, *Running Vacuum in the Universe: current phenomenological status*, arXiv:1601.01668 [gr-qc].
[41] A. Albrecht and J. Magueijo, *A Time varying speed of light as a solution to cosmological puzzles*, Phys. Rev. D **59**, 043516 (1999) doi:10.1103/PhysRevD.59.043516 [astro-ph/9811018].
[42] J. D. Barrow, *Cosmologies with varying light speed*, astro-ph/9811022.
[43] A. Albrecht, *Cosmology with a time varying speed of light*, AIP Conf. Proc. **478**, 263 (1999) doi:10.1063/1.59450 [astro-ph/9904185].
[44] J. W. Moffat, *Variable speed of light cosmology: An Alternative to inflation*, hep-th/0208122.
[45] J. Z. Qi, M. J. Zhang and W. B. Liu, *Observational constraint on the varying speed of light theory*, Phys. Rev. D **90**, no. 6, 063526 (2014) doi:10.1103/PhysRevD.90.063526 [arXiv:1407.1265 [gr-qc]].
[46] S. Landau, P. D. Sisterna and H. Vucetich, *Charge conservation and time varying speed of light*, Phys. Rev. D **63**, 081303 (2001) doi:10.1103/PhysRevD.63.081303 [astro-ph/0007108].
[47] J. Magueijo, *New varying speed of light theories*, Rept. Prog. Phys. **66**, 2025 (2003) doi:10.1088/0034-4885/66/11/R04 [astro-ph/0305457].
[48] M. P. Dabrowski, V. Salzano, A. Balcerzak and R. Lazkoz, *New tests of variability of the speed of light*, EPJ Web Conf. **126**, 04012 (2016). doi:10.1051/epjconf/201612604012
[49] S. Tsujikawa, *Recent status of dark energy*, Mod. Phys. Lett. A **25**, 843 (2010). doi:10.1142/S0217732310000010
[50] L. Amendola and S. Tsujikawa, *Dark Energy: Theory and Observations*, Cambridge University Press, 2010. ISBN: 9780521516006.
[51] M. Li, X. D. Li, S. Wang and Y. Wang, *Dark Energy*, The Universe **1**, no. 4, 24 (2013).
[52] M. Li, X. D. Li, S. Wang and Y. Wang, *Dark Energy: A Brief Review*, Front. Phys. (Beijing) **8**, 828 (2013) doi:10.1007/s11467-013-0300-5 [arXiv:1209.0922 [astro-ph.CO]].
[53] J. D. Barrow and J. Magueijo, *Varying alpha theories and solutions to the cosmological problems*, Phys. Lett. B **443**, 104 (1998) doi:10.1016/S0370-2693(98)01294-5 [astro-ph/9811072].
[54] J. D. Barrow and J. Magueijo, *Can a changing alpha explain the supernovae results?*, Astrophys. J. **532**, L87 (2000) doi:10.1086/312572 [astro-ph/9907354].
[55] J. W. Moffat, *A Model of varying fine structure constant and varying speed of light*, astro-ph/0109350.
[56] M. G. Kozlov, S. G. Porsev, S. A. Levshakov, D. Reimers and P. Molaro, *Mid- and far-infrared fine-structure line sensitivities to hypothetical variability of the fine-structure constant*, Phys. Rev. A **77**, 032119 (2008) doi:10.1103/PhysRevA.77.032119 [arXiv:0802.0269 [astro-ph]].
[57] J. D. Barrow and J. Magueijo, *Local Varying-Alpha Theories*, Mod. Phys. Lett. A **30**, no. 22, 1540029 (2015) doi:10.1142/S0217732315400295 [arXiv:1412.3278 [gr-qc]].
[58] M. Lake and J. Ward, *A Generalisation of the Nielsen-Olesen Vortex: Non-cylindrical strings in a modified Abelian-Higgs model*, JHEP **1104**, 048 (2011) doi:10.1007/JHEP04(2011)048 [arXiv:1009.2104 [hep-ph]].
[59] M. J. Lake, *Cosmic necklaces in string theory and field theory*, Ph.D. Thesis, Queen Mary, University of London (2010).
[60] R. H. Brandenberger and J. Magueijo, *Imaginative cosmology*, hep-ph/9912247.
[61] P. Gopakumar and G. V. Vijayagovindan, *Solutions to cosmological problems with energy conservation and varying c, G and* Λ, Mod. Phys. Lett. A **16**, 957 (2001) doi:10.1142/S0217732301004042 [gr-qc/0003098].
[62] M. Biesiada, *Varying fundamental constants: A dynamical systems approach*, Astrophys. Space Sci. **283**, 511 (2003). doi:10.1023/A:1022564903718
[63] J. D. Barrow, *Cosmological bounds on spatial variations of physical constants*, Phys. Rev. D **71**, 083520 (2005) doi:10.1103/PhysRevD.71.083520 [astro-ph/0503434].
[64] C. G. Scoccola, *Variation of the fine structure constant and the electron mass at early Universe*, arXiv:0906.0329 [astro-ph.CO].
[65] M. L. Yan, *One Electron Atom in Special Relativity with de Sitter Space-Time Symmetry*, Commun. Theor. Phys. **57**, 930 (2012) doi:10.1088/0253-6102/57/6/04 [arXiv:1004.3023 [physics.gen-ph]].
[66] A. Narimani, A. Moss and D. Scott, *Dimensionless cosmology*, Astrophys. Space Sci. **341**, 617 (2012) doi:10.1007/s10509-012-1113-7 [arXiv:1109.0492 [astro-ph.CO]].
[67] L. Hollenstein, R. K. Jain and F. R. Urban, *Cosmological Ohm's law and dynamics of non-minimal electromagnetism*, JCAP **1301**, 013 (2013) doi:10.1088/1475-7516/2013/01/013 [arXiv:1208.6547 [astro-ph.CO]].
[68] S. Pandolfi, *Dark Energy coupling with electromagnetism as seen from future low-medium redshift probes*, J. Phys. Conf. Ser. **566**, no. 1, 012005 (2014). doi:10.1088/1742-6596/566/1/012005
[69] C. J. A. P. Martins, A. M. M. Pinho, R. F. C. Alves, M. Pino, C. I. S. A. Rocha and M. von Wietersheim, *Dark energy and Equivalence Principle constraints from astrophysical tests of the stability of the fine-structure constant*, JCAP **1508**, no. 08, 047 (2015) doi:10.1088/1475-7516/2015/08/047 [arXiv:1508.06157 [astro-ph.CO]].

[70] H. Wei, X. B. Zou, H. Y. Li and D. Z. Xue, *Cosmological Constant, Fine Structure Constant and Beyond*, Eur. Phys. J. C **77**, no. 1, 14 (2017) doi:10.1140/epjc/s10052-016-4581-z [arXiv:1605.04571 [gr-qc]].

[71] H. Wei and D. Z. Xue, *Observational Constraints on Varying Alpha in* $\Lambda(\alpha)CDM$ *Cosmology*,' arXiv:1706.04063 [gr-qc].

[72] G. A. Mena Marugan and S. Carneiro, *Holography and the large number hypothesis*, Phys. Rev. D **65**, 087303 (2002) doi:10.1103/PhysRevD.65.087303 [gr-qc/0111034].

[73] S. Funkhouser, *The Large number coincidence, the cosmic coincidence and the critical acceleration*, Proc. Roy. Soc. Lond. A **462**, 3657 (2006).

[74] P. Burikham, R. Dhanawittayapol and T. Wuthicharn, *A new mass scale, implications on black hole evaporation and holography*, Int. J. Mod. Phys. **31**, no. 16, 1650089 (2016).

[75] C. Bambi, *A Revision of the Generalized Uncertainty Principle*, Class. Quant. Grav. **25**, 105003 (2008) doi:10.1088/0264-9381/25/10/105003 [arXiv:0804.4746 [gr-qc]].

[76] L. Perivolaropoulos, *Cosmological Horizons, Uncertainty Principle and Maximum Length Quantum Mechanics*, arXiv:1704.05681 [gr-qc].

[77] B. Carr, L. Modesto and I. Premont-Schwarz, *Generalized Uncertainty Principle and Self-dual Black Holes*, arXiv: 1107.0708 [gr-qc].

[78] B. Carr, J. Mureika and P. Nicolini, *Sub-Planckian black holes and the Generalized Uncertainty Principle*, JHEP 07, 052 (2016), arXiv:1504.07637.

[79] B. Carr, *The Black Hole Uncertainty Principle Correspondence*, In 1st Karl Schwarzschild Meeting on Gravitational Physics, eds. P. Nicolini, M. Kaminski, J. Mureika, M. Bleicher, pp 159-167 (Springer), arXiv:1402:1427 (2015).

[80] M. J. Lake and B. Carr, *The Compton-Schwarzschild correspondence from extended de Broglie relations*, JHEP 1511, 105 (2015), doi:10.1007/JHEP11(2015)105 arXiv:1505.06994 [gr-qc].

[81] M. J. Lake, *Which quantum theory must be reconciled with gravity? (And what does it mean for black holes?)*, Universe **2**, no. 4, 24 (2016) doi:10.3390/universe2040024 [arXiv:1607.03689 [gr-qc]].

[82] M. J. Lake and B. Carr, *The Compton-Schwarzschild relations in higher dimensions*, arXiv:1611.01913 [gr-qc].

[83] T. P. Singh, *A new length scale for quantum gravity*, arXiv:1704.00747 [gr-qc].

[84] T. P. Singh, *A new length scale, and modified Einstein-Cartan-Dirac equations for a point mass*, arXiv:1705.05330 [gr-qc].

[85] J. Preskill, *Do black holes destroy information?*,' In *Houston 1992, Proceedings, Black holes, membranes, wormholes and superstrings* 22-39, and Caltech Pasadena - CALT-68-1819 (92,rec.Oct.) 17 p [hep-th/9209058].

[86] S. B. Giddings, *The Black hole information paradox*, hep-th/9508151.

[87] S. W. Hawking, *Information loss in black holes*, Phys. Rev. D **72**, 084013 (2005) doi:10.1103/PhysRevD.72.084013 [hep-th/0507171].

IOP Conf. Series: Journal of Physics: Conf. Series **883** (2017) 012002 doi:10.1088/1742-6596/883/1/012002

Generalized Three-Form Field

Pitayuth Wongjun[1,2]

[1] The institute for fundamental study, Naresuan University, Phitsanulok 65000, Thailand

[2] Thailand Center of Excellence in Physics, Ministry of Education, Bangkok 10400, Thailand

E-mail: pitayuthw@nu.ac.th

Abstract. A generalized three-form field is an extended version of the canonical three-form field by considering a Lagrangian of the generalized three-form field as a function of the kinetic and the mass terms. In this work, we investigated cosmological models due to this generalized three-form field. It is found that one can use the three-form field to interpret the non-relativistic matter without the caustic problem. Moreover, by analyzing the dynamical system, a viable model of dark energy due to the generalized three-form field can be obtained.

1. Introduction

It is well-known that there are two accelerated expansion phases of the Universe: the inflationary phase, taking place during the very early era of the Universe, and the dark energy phase, corresponding to the expansion nowadays. A scalar field model is one of simple models to describe these phenomena [1, 2, 3, 4]. Beside the cosmological models due to the scalar field, a three-form field can be successfully used to describe both inflationary phase and dark energy phase [5, 6, 7, 8, 9, 10, 11, 12, 13, 14, 15, 16, 17, 18, 19, 20]. At the perturbation level, it is obvious to see that the three-form field can generate intrinsic vector perturbations while it is not possible for the scalar field. In the present work, by mimicking the k-essence scalar field, we investigate a generalized version of the three-form field by considering a Lagrangian as a function of kinetic and mass terms [18]. We briefly review the important ingredients of the three-form field including energy momentum tensor and the equation of motion in covariant form. The cosmological solution of the model is investigated. A constant equation of state parameter yields the power law of both kinetic and mass terms in the Lagrangian so that one can use the three-form field to interpret the non-relativistic matter without caustic problem which is a problem found in k-essence scalar field model. This issue is discussed in section 2 in the present work. In section 3, we consider the non-constant case of the equation of state parameter and then use the dynamical system to analyze the behavior of the Universe. We found that it is possible to obtain a viable model of dark energy due to the generalized three-form field. Finally, the results are summarized and discussed in section 4.

2. Generalized Three-Form Filed

In this section, we will review the generalized three-form field model by considering a Lagrangian of the three-form field, $A_{\alpha\beta\gamma}$, as a function of the kinetic and mass terms as follows, [18]

$$S = \int d^4x\sqrt{-g}\left[\frac{M_{Pl}^2}{2}R + P(K, y)\right], \tag{1}$$

where the kinetic term and scalar quantity of the three-form field are expressed as

$$K = -\frac{1}{48}F_{\alpha\beta\gamma\delta}F^{\alpha\beta\gamma\delta}, \quad y = \frac{1}{12}A_{\alpha\beta\gamma}A^{\alpha\beta\gamma}, \quad F_{\mu\nu\rho\sigma} = \nabla_{[\mu}A_{\nu\rho\sigma]}. \tag{2}$$

From this action, the equations of motion and the energy momentum tensor of the three-form can be written as

$$E_{\alpha\beta\gamma} = \nabla_\mu\left(P_{,K}F^{\mu}{}_{\alpha\beta\gamma}\right) + P_{,y}A_{\alpha\beta\gamma} = 0, \tag{3}$$

$$T_{\mu\nu} = \frac{1}{6}P_{,K}F_{\mu\rho\sigma\alpha}F_{\nu}{}^{\rho\sigma\alpha} - \frac{1}{2}P_{,y}A_{\mu\rho\sigma}A_{\nu}{}^{\rho\sigma} + Pg_{\mu\nu}. \tag{4}$$

where the notation with subscript $P_{,x}$ denotes $P_{,x} = \partial_x P$. The energy momentum tensor is conserved up to the equation of motion as $6\nabla_\mu T^\mu_\nu = F_{\nu\alpha\beta\gamma}E^{\alpha\beta\gamma} = 0$.

By considering a flat Friedmann-Lemaitre-Robertson-Walker (FLRW) metric, the components of the three-form field, $A_{\alpha\beta\gamma}$, can be written as $A_{0ij} = 0, A_{ijk} = a^3\varepsilon_{ijk}X(t)$ where ε_{ijk} is the three-dimensional Levi-Civita symbol with $\varepsilon_{123} = 1$. As a result, the energy density and pressure of the three-form field can be expressed as

$$\rho_X = 2KP_{,K} - P, \tag{5}$$

$$p_X = -\rho_X - 2yP_{,y}. \tag{6}$$

For the equation of state parameter, $w_X = p_X/\rho_X = -1 - 2yP_{,y}/\rho_X$, a partial differential equation of P can be written as

$$2yP_{,y} + (1 + w_X)2KP_{,K} = (1 + w_X)P. \tag{7}$$

By taking the equation of state parameter to be a constant, one can solve this equation to obtained the exact solution as $P = P_0K^\nu y^\mu$, where P_0 is an integration constant and μ, ν are the exponent constants. As a result, the equation of state parameter can be expressed in terms of the constants as

$$w_X = -1 + \frac{2\mu}{1-2\nu}, \quad \nu \neq \frac{1}{2}. \tag{8}$$

From this expression, the description of non-relativistic matter can be obtained by choosing a proper parameters to yield $w_X = 0$, for example, $\mu = 1, \nu = -1/2$. Therefore, the Lagrangian of the three-form is still finite and then the model can avoid the caustic problem which is found in the k-essence scalar field case [18]. Moreover, one can relax the assumption of constant w by assuming $w_X = w_X(y)$. As a result, one of the solutions can be written as $P = P_0K^\nu e^{\frac{(1-2\nu)}{2}\lambda y}$ [18] where $w_X = -1 + \lambda y$ and λ is a constant. From this solution, we found that it is possible to obtain the solution which admits the dark energy model. We will investigate this possibility in detail by using the dynamical system in the next section.

3. Dynamical system of the dark energy model

In this section, the dynamics of the Universe will be explored by using the dynamical system. By including the non-relativistic matter and radiation into the generalized three-form field model in Eq. (1), non-zero components of Einstein equation can be written as

$$1 = \Omega_X + \Omega_m + \Omega_r, \tag{9}$$

$$\frac{2H'}{3H^2} = -1 - w_X\Omega_X - \frac{1}{3}\Omega_r, \tag{10}$$

where Ω is the density parameter, $H = \dot{a}/a$ is the Hubble parameter, prime denotes the derivative with respect to $N = \ln a$ and the subscripts m and r represent the quantity corresponding to matter and radiation respectively. The autonomous system is obtained by using the conservation of energy momentum tensor for each species together with the Einstein equations, Eq. (9) and Eq. (10). As a result, the autonomous system corresponding to the generalized three-form field can be written as

$$\Omega'_X = -3\Omega_X\left(w_X(1-\Omega_X) - \frac{1}{3}(1-\Omega_X-\Omega_m)\right), \quad (11)$$

$$\Omega'_m = 3\Omega_m\left(w_X\Omega_X + \frac{1}{3}(1-\Omega_X-\Omega_m)\right), \quad (12)$$

$$w'_X = (1+w_X)\Gamma\frac{y'}{y}, \quad \Gamma = 1 - \frac{yQ_{,y}}{Q} + \frac{yQ_{,yy}}{Q_{,y}}, \quad (13)$$

where we have chosen $P = P_0K^\nu Q(y)$. From this system, one can see that the equation of w_X is decoupled from the others. Note that $\Gamma = 0$ for $Q \propto y^\mu$ which implies that $w'_X = 0$ as we expect for the case of constant w_X. From the autonomous system, there are three fixed points as follows; matter-dominated fixed point $(\Omega_X, \Omega_m, \Omega_r) = (0,1,0)$, radiation-dominated fixed point $(\Omega_X, \Omega_m, \Omega_r) = (0,0,1)$ and dark-energy-dominated fixed point $(\Omega_X, \Omega_m, \Omega_r) = (1,0,0)$. Considering equation of w_X in Eq. (13), it is convenient to choose the solution such that $y'/y = \alpha$ where α is a constant. This choosing is qualitatively similar to one in usual three-form field model [10, 11, 12]. Therefore, the equation for w_X in Eq. (13) has a fixed point at $w_X = -1$ and its dynamics and stability depend on the sign of $\alpha\Gamma$. If $\alpha\Gamma < 0$, where $w_X = -1$, the dark-energy-dominated fixed point will be a stable fixed point. By choosing Q to be a more general function than one in [18] as $Q = \exp\left(\frac{(1-2\nu)}{2}\lambda y^n\right)$, we obtain $\Gamma = n$. The conditions to avoid ghost and Laplacian instabilities have been investigated in [18] and then provide the conditions to the model parameter as $n(2\nu - 1) > 0$ and $n(2 + \lambda y^n) > 1$. The sufficient conditions for healthy theory are $n \geq 1/2, \nu > 1/2, \lambda > 0$. Hence, Γ must be always positive to avoid the instabilities. In order to obtain a viable model of dark energy, α must be negative. Moreover, the value of $|\alpha|$ must be sufficiently small to guarantee that w_X does not deviate too much over the history of the Universe. We numerically show this viable model by setting $\alpha = -0.1$ and $n = 1$ as found in Fig 1. Note that for $n \geq 2$, the result is not significantly different as long as $|\alpha|$ is sufficiently small enough.

4. Summary

In this paper, we investigated the generalized three-form field by considering a more general Lagrangian as $L = P(K, y)$ [18] where K is a canonical kinetic term and y is a standard mass term of the three-form field. From the energy momentum tensor of the three-form field, we found the relation between P and its partial derivatives via the equation of state parameter, w_X. By assuming constant w_X, the equation can be exactly solved and it solution can provide the description of non-relativistic matter without caustic problem as found in k-essence scalar field. For non-constant w_X, we specialize our attention in the case $w_X = w_X(y)$ and a suitable Lagrangian is chosen as $P = P_0K^\nu \exp\left(\frac{(1-2\nu)}{2}\lambda y^n\right)$. By analyzing dynamical system, we found that it is possible to obtain a viable model of dark energy due to the generalized three-form field. We also confirm our analysis by using numerical method which can be found in Fig. 1.

Even though the model can be used to explain the cosmic acceleration of the Universe, it cannot provide the way to solve the coincidence problem. It is of interest to investigate dark energy and dark matter coupling to alleviate the coincidence problem in which both contents are constructed from the three-form fields. We leave this investigation for further work.

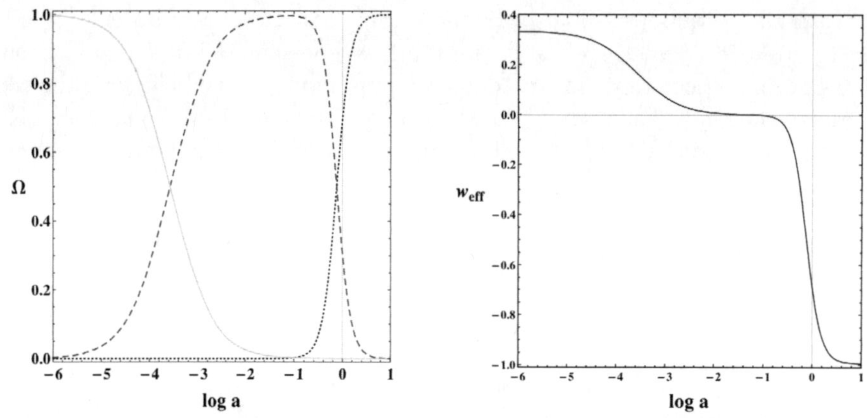

Figure 1. The left panel shows the evolution of the density parameters for three-form dark energy (black dotted line), matter (blue dashed line) and radiation (green solid line). The right panel shows the evolution of the effective equation of state parameter w_{eff}. The parameters are set as $\alpha = -0.1$ and $n = 1$.

Acknowledgments

The author is supported by Thailand Research Fund (TRF) through grant TRG5780046. The author would like to thank Lunchakorn Tannukij for value discussion and comments.

References

[1] Armendariz-Picon C, Mukhanov V F and Steinhardt P J 2000 A Dynamical solution to the problem of a small cosmological constant and late time cosmic acceleration *Phys. Rev. Lett.* **85** 4438
[2] Armendariz-Picon C, Mukhanov V F and Steinhardt P J 2001 Essentials of k essence *Phys. Rev.* D **63** 103510
[3] Chiba T, Okabe T and Yamaguchi M 2000 Kinetically driven quintessence *Phys. Rev.* D **62** 023511
[4] Boubekeur L, Creminelli P, Norena J and Vernizzi F 2008 Action approach to cosmological perturbations: the 2nd order metric in matter dominance *JCAP* **0808** 028
[5] Germani C and Kehagias A 2009 P-nflation: generating cosmic Inflation with p-forms *JCAP* **0903**, 028
[6] Koivisto T S, Mota D F and Pitrou C 2009 Inflation from N-Forms and its stability *JHEP* **0909**, 092 (2009)
[7] Kobayashi T and Yokoyama S 2009 Gravitational waves from p-form inflation *JCAP* **0905**, 004
[8] Germani C and Kehagias A 2009 Scalar perturbations in p-nflation: the 3-form case *JCAP* **0911**, 005
[9] Koivisto T S and Nunes N J 2009 Inflation and dark energy from three-forms *Phys. Rev.* D **80**, 103509
[10] Koivisto T S and Nunes N J 2010 Three-form cosmology *Phys. Lett.* B **685**, 105 (2010)
[11] Ngampitipan T and Wongjun P 2011 Dynamics of three-form dark energy with dark matter couplings *JCAP* **1111**, 036
[12] De Felice A, Karwan K and Wongjun P 2012 Stability of the 3-form field during inflation *Phys. Rev.* D **85**, 123545
[13] Koivisto T S and Nunes N J 2013 Coupled three-form dark energy *Phys. Rev.* D **88**, 123512
[14] Kumar K S, Marto J, Nunes N J and Moniz P V 2014 Inflation in a two 3-form fields scenario *JCAP* **1406**, 064
[15] Barros B J and Nunes N J 2016 Three-form inflation in type II Randall-Sundrum *Phys. Rev.* D **93**, no. 4, 043512
[16] De Felice A, Karwan K and Wongjun P 2012 Reheating in 3-form inflation *Phys. Rev.* D **86**, 103526
[17] Koivisto T S and Urban F R 2012 Three-magnetic fields *Phys. Rev.* D **85**, 083508
[18] Wongjun P 2016 A Perfect Fluid in Lagrangian Formulation due to Generalized Three-Form Field (*Preprint* arXiv:1602.00682 [gr-qc])
[19] Sravan Kumar K, Mulryne D J, Nunes N J, Marto J and Vargas Moniz P 2016 Non-Gaussianity in multiple three-form field inflation *Phys. Rev.* D **94**, no. 10, 103504
[20] Morais J, Bouhmadi-Lopez M, Sravan Kumar K, Marto J and Tavakoli Y 2017 Interacting 3-form dark energy models: distinguishing interactions and avoiding the Little Sibling of the Big Rip *Phys. Dark Univ.* **15** 7

IOP Conf. Series: Journal of Physics: Conf. Series **883** (2017) 012003 doi:10.1088/1742-6596/883/1/012003

Time-dependent Lagrangian Perturbation Theory with The Dynamical Dark Energy

Teeraparb Chantavat[1]

[1] ThEP's Laboratory of Cosmology and Gravity, The Institute for Fundamental Study "The Tah Poe Academia Institute", Naresuan University, Phitsanulok, 65000, Thailand

E-mail: teeraparbc@nu.ac.th

Abstract. This article gives a brief overview of the time-dependent Lagrangian perturbation theory as a theoretical tool to predict the behavior of galaxy clustering in quasi-linear regimes. While most of the Lagrangian perturbation theories are based on an Einstein-De Sitter space where the growth function is time independent, more accurate theoretical predictions could be achieved by time-dependent growth functions in a more general cosmology. This is directly applicable to the dynamical dark-energy where the clustering behavior can be recognized with time-dependent growth functions.

1. Introduction

One of the main goals of cosmology is to understand the large-scale structure of the Universe. The structures that we see in the galaxy surveys today are derived from the widely accepted gravitational instability with primordial matter density fluctuations [1, 2]. A quantitative understanding of the dynamics of the structure formation requires a theoretical modeling that can be applied with statistical tools to test the theory against observations. Predicting the behavior of the formation of the structures is crucially important as a direct way to validate the current dynamical theory of gravity with the observations such as the Baryonic Acoustic Oscillations (BAOs) [3]. As a simple, but accurate, model for numerical predictions, an irrotational and pressureless fluid of cold dark matter physics is normally considered in the non-relativistic regime [4]. The observable could, in principle, be predicted. When higher-order terms are considered, one would expect that the numerical predictions should be more accurate; however, with the cost of higher complexities and computational time.

There are two main classes of non-relativistic cosmological perturbation theory. One is expressed observable quantities in terms of the Eulerian frame, called Standard Perturbation Theory (SPT) [5, 6, 7]. Other one is given in the Lagrangian frame, call Lagrangian Perturbation Theory (LPT) [8, 9]. This *article* I shall give a brief overview of the time dependent Lagrangian perturbation theory up to the third order with is sufficiently high enough for accurate numerical predictions especially the applications of the dynamical dark energy where the equation of state is a function of time.

2. Lagrangian Perturbation Theory

In Lagrangian perturbation theory, the dynamical variable is the displacement field $\boldsymbol{x}(t) = \boldsymbol{q}(\boldsymbol{x}) + \boldsymbol{\Psi}(\boldsymbol{q}, t)$, where $\boldsymbol{\Psi}$ is the displacement field, $\boldsymbol{q}$ is the position of the particle at the initial time. The dynamical equation is given by

$$J(\boldsymbol{q},t)\nabla\cdot\left[\frac{\mathrm{d}^2}{\mathrm{d}t^2}\boldsymbol{\Psi}+2H(t)\frac{\mathrm{d}}{\mathrm{d}t}\boldsymbol{\Psi}\right]=\frac{3}{2}\Omega_M(t)H^2(t)(J-1), \tag{1}$$

where $J(\boldsymbol{q},t)$ is the Jacobian of the transformation between $\boldsymbol{x}$ and $\boldsymbol{q}$. The Hubble parameter, $H(t)$, composes of the component of the universe mainly cold dark matter $\Omega_M(t)$ and dynamical dark energy $\Omega_{\mathrm{DE}}(t)$ with the constant curvature parameter k.

$$\frac{H^2(t)}{H_0^2}=\Omega_M(t)+\Omega_{\mathrm{DE}}(t)-\frac{kc^2}{a^2(t)}. \tag{2}$$

The dark energy depends on the equation of state parameter as

$$w(t)=w_0+w_1\frac{z(t)}{1+z(t)}, \tag{3}$$

where $z(t)$ is the redshift w_0 and w_1 are constants [10, 11].

3. The Evolution Equations for Perturbations

In order to describe the large scale structure that we see in the universe, we shall expand the density field and the velocity field as

$$\delta(\boldsymbol{x},t) = \delta_1(\boldsymbol{x},t)+\delta_2(\boldsymbol{x},t)+\ldots, \tag{4}$$
$$\boldsymbol{v}(\boldsymbol{x},t) = \boldsymbol{v}_1(\boldsymbol{x},t)+\boldsymbol{v}_2(\boldsymbol{x},t)+\ldots, \tag{5}$$

where δ_n and $\boldsymbol{v}_n$ is the corresponding nth order perturbative overdensity and velocity field respectively.

3.1. First Order

To the first order, we can write the overdensity as

$$\delta_1(\boldsymbol{x},t)=D_1(t)\epsilon, \tag{6}$$

where $D_1(t)$ is the linear growth function. $\epsilon(\boldsymbol{x})$ is the initial background overdensity. The linear growth function can be found by solving the second-order differential equation with appropriated initial conditions;

$$\ddot{D}_1+2H\dot{D}_1-\frac{3}{2}\Omega_M H^2 D_1=0. \tag{7}$$

3.2. Second Order

In order to calculate the second order term, the velocity potential, $\varphi(\boldsymbol{x})$, is needed,

$$\boldsymbol{v}_1(\boldsymbol{x},t)=-\dot{D}_1\boldsymbol{\nabla}\varphi(\boldsymbol{x}). \tag{8}$$

It turns out that the time-dependent growth function for the second-order perturbation $D_2(t)$ is given by

$$\ddot{D}_2+2H\dot{D}_2-\frac{3}{2}\Omega_M H^2 D_2=\frac{3}{2}\Omega_M H^2 D_1^2. \tag{9}$$

As one can see the second order growth function is proportional to D_1^2.

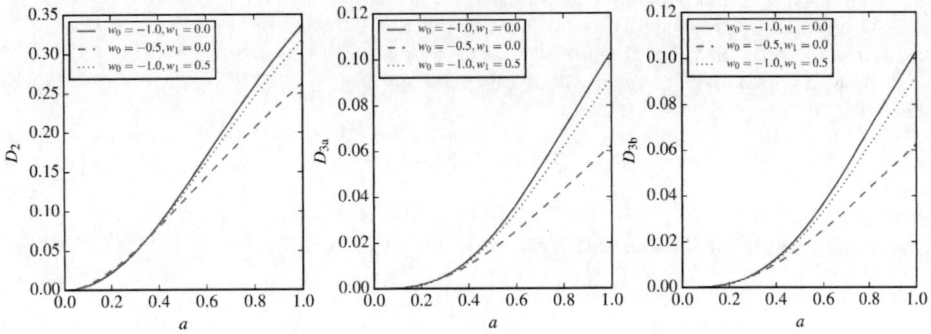

Figure 1. The time evolution functions for third order terms $\Omega_{M,0} = 0.3$, $\Omega_{\rm DE,0} = 0.7$ for different values of w_0 and w_1 with the normalization condition $D_1 = 1$ at $a = 1$.

3.3. Third Order

There are two time dependent growth functions for the third order terms $D_{3\rm a}(t)$ and $D_{3\rm b}(t)$ which are given by

$$\ddot{D}_{3\rm a} + 2H\dot{D}_{3\rm a} - \frac{3}{2}\Omega_M H^2 D_{3\rm a} = \frac{3}{2}\Omega_M H^2 \left(D_1^3 + 3D_1 D_2\right), \tag{10}$$

$$\ddot{D}_{3\rm b} + 2H\dot{D}_{3\rm b} - \frac{3}{2}\Omega_M H^2 D_{3\rm b} = \frac{3}{2}\Omega_M H^2 D_1^3. \tag{11}$$

Similar, to the second-order growth function, the third-order growth functions are proportional to D_1^3.

4. Results

In this article, we shall assume a flat cosmology with dark energy component i.e. $\Omega_{\rm M,0} = 0.3, \Omega_{\rm DE,0} = 0.7$ and $k = 0$. The time evolution equations are solved using fourth order Runge-Kutta method with the condition $D_1(a = 1) = 1$ i.e. the linear growth function is equal to unity at the present epoch. The second and third-order growth functions are shown in FIG. 1 for different cosmological models with different value of w_0 and w_1. The growth functions are sensitive to the equation of state parameter of the dynamical dark energy. The difference in growth rate is more pronounced at the present epoch.

5. Conclusions

We study the third order Lagrangian perturbation theory and derive the time dependent evolution functions. We find that there are different growth functions for different perturbative orders; however, there are two independent evolution functions for the third-order quantities which describe different terms in the expansion. The evolution function is sensitive to the equation of state parameter of the dynamical dark energy w which is helpful for large-scale structure surveys, where the effect of time-dependent w would be discernible. While the effect of w on large-scale structure would look marginally different from the time-independent one's, it is possible to stack the differences in many redshift slides to amplify the difference. This can be achieved with the the Alcock-Paczyński (AP) test [12] where the change in the angular diameter distance can be recognized. Our approach is advantageous to the standard perturbation theory (SPT) approach since we only need to specify the initial conditions for the first order quantities in the density and velocity field.

References

[1] Peebles P J E 1980 *The large-scale structure of the universe*
[2] Peebles P J E 1993 *Principles of Physical Cosmology*
[3] Eisenstein D J and Hu W 1998 *ApJ* **496** 605–614 (*Preprint* astro-ph/9709112)

[4] Carlson J, White M and Padmanabhan N 2009 *Physical Review D* **80** 043531 (*Preprint* 0905.0479)
[5] Goroff M H, Grinstein B, Rey S J and Wise M B 1986 *ApJ* **311** 6–14
[6] Jain B and Bertschinger E 1994 *ApJ* **431** 495–505 (*Preprint* astro-ph/9311070)
[7] Scoccimarro R and Frieman J 1996 *ApJS* **105** 37 (*Preprint* astro-ph/9509047)
[8] Zel'dovich Y B 1970 *A&A* **5** 84–89
[9] Bouchet F R, Colombi S, Hivon E and Juszkiewicz R 1995 *A&A* **296** 575 (*Preprint* astro-ph/9406013)
[10] Chevallier M and Polarski D 2001 *International Journal of Modern Physics D* **10** 213–223 (*Preprint* gr-qc/0009008)
[11] Linder E V 2003 *Physical Review Letters* **90** 091301 (*Preprint* astro-ph/0208512)
[12] Alcock C and Paczynski B 1979 *Nature* **281** 358

IOP Conf. Series: Journal of Physics: Conf. Series **883** (2017) 012004 doi:10.1088/1742-6596/883/1/012004

Searching for modified gravity: a conformal sector?

J Ovalle[1,2], R Casadio[3,4] and A Sotomayor[5]

[1] Departamento de Física, Universidad Simón Bolívar, AP 89000, Caracas 1080A, Venezuela
[2] The Institute for Fundamental Study, Naresuan University, Phitsanulok 65000, Thailand
[3] Dipartimento di Fisica e Astronomia, Alma Mater Università di Bologna, via Irnerio 46, 40126 Bologna, Italy
[4] Istituto Nazionale di Fisica Nucleare, Sezione di Bologna, I.S. FLAG, viale Berti Pichat 6/2, 40127 Bologna, Italy
[5] Departamento de Matemáticas, Universidad de Antofagasta, Antofagasta, Chile

E-mail: jovalle@usb.ve

E-mail: casadio@bo.infn.it

E-mail: adrian.sotomayor@uantof.cl

Abstract. We conjecture that any modification of general relativity can be studied by the minimal geometric deformation approach provided that such modification can be represented by a traceless energy-momentum tensor.

1. Introduction

As is well known, there are some fundamental questions associated with the gravitational interaction which general relativity (GR) cannot answer satisfactorily, namely the dark matter problem and the introduction of dark energy to explain the accelerated expansion of the universe. Moreover, the incompatibility between GR and the Standard Model of particle physics, or equivalently, the failure to quantize GR by the same successful scheme used with the other fundamental interactions. Such issues have motivated the search for new gravitational theories beyond GR that could help to explain part of these problems. Among these alternatives, we have extra-dimensional theories. In this context the minimal geometric deformation approach (MGD), originally proposed [1] in the the Randall-Sundrum brane-world [2, 3], was developed and later extended to investigate new black hole solutions [4, 5] (For a recent brief review, see [7]). While the brane-world is still an attractive scenario, to find interior solutions for self-gravitating systems is a difficult task, mainly due to the existence of non-linear terms in the matter fields. Despite these complications, the MGD has proven to be useful, among other things, to derive exact and physically acceptable solutions for spherically symmetric and non-uniform stellar distributions [6, 10] as well; to express the tidal charge in the metric found in Ref. [11] in terms of the usual Arnowitt-Deser-Misner (ADM) mass [12]; to study microscopic black holes [13]; to clarify the role of exterior Weyl stresses acting on compact stellar distributions [14, 15]; to extend the concept of variable tension introduced in Refs. [16] by analyzing the shape of the black string in the extra dimension [17]; to prove, contrary to previous claims, the consistency of a Schwarzschild exterior [18] for a spherically symmetric self-gravitating system made of regular matter in the brane-world; to derive bounds on extra-dimensional parameters [19] from the observational results of the classical tests of GR in the Solar system; to investigate the gravitational lensing phenomena beyond GR [20]; as well as to determine the critical stability region for Bose-Einstein condensates in gravitational systems [21].

2. The MGD: a qualitative approach

Let us start by assuming that there is a "superior" gravitational theory, superior in the sense that it could explain the dark matter problem, what dark energy is, and maybe to explain the quantum problem. Of course, this new gravitational theory must contain GR at some sector, that is, it should be capable of reproducing GR at solar system scale for instance. The reason for this is the fact that GR is a well tested and successful theory, at least at solar system scale. This new gravitational theory can be seen as GR plus some "corrections" These corrections always produce consequences on the energy-momentum tensor $T_{\mu\nu}$. Indeed, these corrections can be consolidated as part of an effective energy-momentum tensor $\theta_{\mu\nu}$ in such a way that

$$T_{\mu\nu} \rightarrow T_{\mu\nu} + \alpha\theta_{\mu\nu} \ , \tag{1}$$

where α is a free parameter associated with the new gravitational sector not described by GR. Of course, when we turn off this free parameter we return to the domain of GR. This limit represents not only a critical point for a consistent extension of GR, but also a non-trivial problem that must be treated carefully. In this respect, one cannot try and change GR without considering the well-established and very useful Lovelock's theorem [22], which severely restricts any possible ways of modifying GR in four dimensions. We will now show the simplest generic way.

The simplest way to extend GR is by considering a modified Einstein-Hilbert action,

$$S = \int \left[\frac{R}{2\,k^2} + \mathcal{L}\right]\sqrt{-g}\,d^4x + \alpha\,(\text{correction})\ , \tag{2}$$

where the generic correction shown in Eq. (2) should be, of course, a well justified and physically motivated expression. At this stage the GR limit, obtained by setting $\alpha = 0$, is just a trivial issue, so everything looks consistent. Indeed, we may go further and calculate the equations of motion by the usual way, hence

$$R_{\mu\nu} - \frac{1}{2}\,R\,g_{\mu\nu} = k^2\,(T_{\mu\nu} + \alpha\theta_{\mu\nu})\ . \tag{3}$$

The new term $\theta_{\mu\nu}$ in Eq. (3) in general has new fields, like scalar, vector and tensor fields, all of them coming from the new gravitational sector not described by Einstein's theory. At this stage the GR limit, again, is a trivial issue, since $\alpha = 0$ leads to the standard Einstein's equations $G_{\mu\nu} = k^2\,T_{\mu\nu}$. However, when the system of equations given by the expression (3) is solved, in general the solution eventually found cannot reproduce the GR limit by simply setting $\alpha = 0$. The cause of this problem is the non-linearity of Eq. (3), and should not be a surprise. To clarify this point, let us consider a generic spherically symmetric solution for the radial metric component, namely

$$g_{rr}^{-1} = 1 - \frac{2\,m(r)}{r}\ , \tag{4}$$

where m is the mass function of the self-gravitating system. Now let us consider the effect of the new gravitational sector in (2), which modifies the equation of motion in Eq. (3) by the new term $\theta_{\mu\nu}$. This new term produces a "geometric deformation" on the radial metric component (4), which generically may be written as

$$g_{rr}^{-1} = 1 - \frac{2\,m(r)}{r} + (\text{geometric deformation})\ , \tag{5}$$

IOP Conf. Series: Journal of Physics: Conf. Series **883** (2017) 012004 doi:10.1088/1742-6596/883/1/012004

where by *geometric deformation* one should understand a deviation from the generic GR solution in (4). It is now very important to note that the deformation (5) always takes the form

$$(\text{geometric deformation}) = X + \alpha Y \ . \tag{6}$$

This expression is very significant, since it shows that the GR limit cannot be *a posteriori* recovered by setting $\alpha = 0$, since there is a "sector" denoted by X which does not depend on α. Consequently, the GR solution is not trivially contained in this extension, and one might say that we have an extension to GR which does not contain GR. A method that solves the non-trivial issue of consistency with GR described above is the so-called *Minimal Geometric Deformation* MGD approach [1]. The idea is to keep under control the anisotropic consequences on GR appearing in the extended theory, in such a way that the α-independent sector in the geometric deformation shown as X in Eq. (6) always vanishes. This is accomplished when a GR solution is forced to remain a solution in the extended theory. Roughly speaking, we need to introduce the GR solution into the new theory, as far as possible. This provides the foundation for the MGD approach.

3. MGD and $f(R)$ theories

We know the MGD works very well to generate new brane-world solutions from already known GR solutions. Since the MGD does not care about the source of the modification, in the sense that these modifications always can be consolidated as an energy-momentum tensor $\theta_{\mu\nu}$, there is no reason to believe it will not work properly in any other extension beyond the brane-world. In this respect, it would be interesting to test this approach in the context of $f(R)$ theories. The reason is that these theories have interesting properties [34, 35, 36, 37] and appears in a natural way in superstring theories.

When we use the MGD in $f(R)$ theories, we found that the geometric deformation $g^*(r)$ undergone by the Schwarzschild metric, namely

$$ds^2 = \left(1 - \frac{2\mathcal{M}}{r}\right) dt^2 - \left(1 - \frac{2\mathcal{M}}{r} + \alpha\, g^*(r)\right)^{-1} dr^2 - d\Omega^2 \ . \tag{7}$$

yields to the trivial solution $g^*(r) = 0$. This is unexpected since we know there are solutions beyond the Schwarzschild metric in the spherically symmetric vacuum under $f(R)$, hence we should at least to reproduce some of these solutions. We can conclude thus that the MGD in its original form and $f(R)$ are incompatible. Therefore we need to modify the MGD approach to make it compatible with $f(R)$, but there are not many options to accomplish this. Indeed, the only possible extension is to introduce by hand a deformation on the temporal metric component, as developed in Ref. [4]. When this extension is used in $f(R)$ we end up with a very complicated system of equations. However, the model $f(R) = R + \alpha R^2$ produces a simpler system which could be useful to find new vacuum solution beyond the Schwarzschild metric. This should not be a surprise, since it is well known that the pure $f(R) = R^2$ is conformally equivalent to Einstein gravity with a minimally coupled scalar field and a cosmological constant [38], so we should anticipate that in this particular case the MGD and $f(R)$ seems to be compatible with the searching for new vacuum solutions. At this stage we can ask ourselves what do the effective Einstein field equations in the brane-world and $f(R) = R^2$ have in common? In the case of the brane-world, if high-energy corrections are removed keeping only non-local corrections, then the energy-momentum tensor representing the new gravitational sector is traceless. Hence it can be associated with a conformal gravitational sector. We know that in this case, namely, the brane-world with only non-local corrections, the MGD works very well. All this tell us that any modification of general relativity can be studied by the MGD provided that such modification can be represented by a traceless energy-momentum tensor, or equivalently, associated with a conformal gravitational sector.

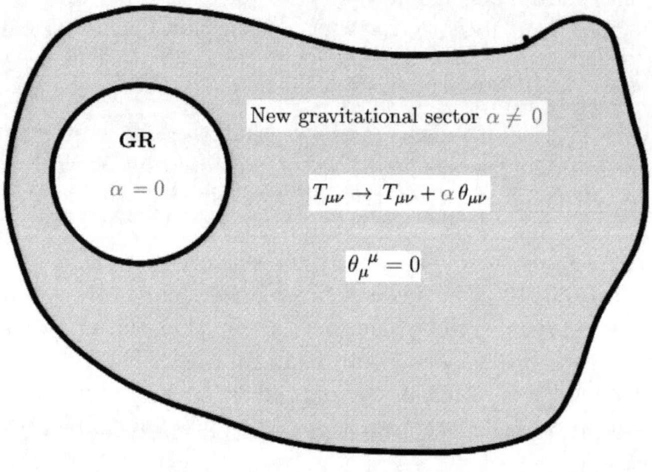

Figure 1. When the energy-momentum tensor $\theta_{\mu\nu}$ associated with the new gravitational sector is traceless, the MGD works properly. It seems to show a link between conformal invariance and the MGD.

Acknowledgements
A.S. is partially supported by Project Fondecyt 1161192, Chile.

References
[1] Ovalle J, 2010 *Proc. ICGA9 Gravitation and Astrophysics* (Wuhan) (Singapore: World Scientific) p 173
[2] Randall L and Sundrum R 1999 *Phys. Rev. Lett.* **83** 3370
[3] Randall L and Sundrum R 1999 *Phys. Rev. Lett.* **83** 4690
[4] Casadio R, Ovalle R and da Rocha R 2015 *Class. Quantum Grav.* **32**, 215020
[5] Ovalle J 2016 *Int. J. Mod. Phys. Conf. Ser.* **41** 1660132
[6] Ovalle J 2008 *Mod. Phys. Lett. A.* **23** 3247
[7] Ovalle J, Casadio R and Sotomayor A, *The Minimal Geometric Deformation Approach: a brief introduction*; arXiv:1612.07926 [gr-qc]
[8] Ovalle J 2009 *Int. J. Mod. Phys. D* **18** 837
[9] Ovalle J 2010 *Mod. Phys. Lett. A* **25** 3323
[10] Ovalle J 2012 *Proceedings of the Twelfth Marcel Grossmann Meeting on General Relativity* (Paris)(Singapore: World Scientific)p 2243
[11] Dadhich N, Maartens R, Papadopoulos P and Rezania V 2000*Phys. Lett. B* **487** 1
[12] Casadio R and Ovalle J 2012 *Phys. Lett. B* **715** 251
[13] Casadio R and Ovalle J 2014 *Gen. Relat. Grav.* **46** 1669
[14] Ovalle J and Linares F 2013 *Phys. Rev. D* **88** 104026
[15] Ovalle J, Linares F, Pasqua A and Sotomayor A 2013 *Class. Quantum Grav.* **30** 175019
[16] Gergely L A 2008 *Phys. Rev. D* 78:084006
[17] Casadio R, Ovalle J and da Rocha R 2014 *Class. Quantum Grav.* **30** 175019
[18] Ovalle J, Gergely L A and Casadio R 2015 *Class. Quantum Grav.* **32** 045015
[19] Casadio R, Ovalle J and da Rocha R 2015 *Europhys. Lett.* **110** 40003
[20] Cavalcanti R T, Goncalves da Silva A and da Rocha R 2016 *Class. Quantum Grav.* **33** 215007
[21] Casadio R and da Rocha R 2016 *Phys. Lett. B* **763** 434
[22] Lovelock D 1971 *J. Math. Phys.* **12** 498
[23] Maartens R 2004 *Living Rev. Rel.* 7
[24] Maartens R and Koyama K 2010 *Living Rev. Relativity* **13** 5
[25] Shiromizu T, Maeda K and Sasaki M 2000 *Phys.Rev. D* **62** 024012
[26] Germani C and Maartens R 2001 *Phys. Rev. D* **64** 124010
[27] Harko T and Lake M J 2014 *Phys. Rev. D* **89** 064038
[28] Linares F, Garcia-Aspeitia M A and Urea-Lopez L A 2015 *Phys. Rev. D* **92** 024037

IOP Conf. Series: Journal of Physics: Conf. Series **883** (2017) 012004 doi:10.1088/1742-6596/883/1/012004

[29] Casadio R and Mazzacurati L 20003 *Mod. Phys. Lett. A* **18** 651
[30] da Rocha R and Hoff da Silva J M 2012 *Phys. Rev. D* **85** 046009
[31] Koyama K and Maartens R 2006 *JCAP* **016** 0601
[32] Viznyuk A and Shtanov Y 2007 *Phys. Rev. D* **76** 064009
[33] Casadio R, Fabbri A and Mazzacurati L 2002 *Phys. Rev. D* **65** 084040
[34] De Felice A and Tsujikawa S 2010 *Living Rev. Rel.* **13** 3
[35] Sotiriou T P and Faraoni V 2010 *Rev. Mod. Phys.* **82** 451
[36] Capozziello A and De Laurentis M 2011 *Phys. Rept.* **509** 167
[37] Capozziello S, Cardone V F and Troisi A 2005, *Phys. Rev. D* **71** 043503
[38] Alvarez-Gaume L, Kehagias A, Kounnas C, Lust D and Riotto A 2016 *Fortsch. Phys.* **64** 176

IOP Conf. Series: Journal of Physics: Conf. Series **883** (2017) 012005 doi:10.1088/1742-6596/883/1/012005

Mass-varying massive gravity with k-essence

Lunchakorn Tannukij[a,b,c]

[a]Department of Physics, Faculty of Science, Mahidol University, Bangkok 10400, Thailand
[b]The Institute for Fundamental Study, Naresuan University, Phitsanulok 65000, Thailand
[c]Department of Physics, Hanyang University, Seoul 133-891, Korea

E-mail: l_tannukij@hotmail.com

Abstract. In this work I review the specific class of mass-varying massive gravity in which a corresponding graviton is given nonzero varying mass. Particularly in this theory, the graviton mass varies which is driven by a rolling k-essence field. Firstly, the model action is expressed along with its equations of motion. It is found from those equations of motion that not only the cosmic accelerating expansion but also the presence of dark matter can be realized in this context. Due to the complexity of its dynamics, the method of dynamical analysis is used to investigate evolutions of the cosmological system. It is found that the corresponding cosmological evolutions can explain the cosmic coincidence problem, where the effective dark matter is presented while the system is in the phase of the cosmic accelerating expansion. I then conclude that the evolutions obtained here can alleviate the cosmic coincidence problem while in some aspects the fine-tuning problem arises.

1. Introduction

In the world of physics, gravity is one of the most interesting topics as being one of the four forces in nature. To gain more understanding on nature, physicists were searching for and formulating theories of gravity for centuries. One of the most successful gravity theory was proposed by Einstein, formally known as general relativity. General relativity has brought useful insights to the gravitational physics especially in the aspects of the solar system, e.g. it has proved itself useful in the context of the Mercury's trajectory. Despite its advantages, the situation is totally different on the larger scales. For the scale of galaxies general relativity fails to describe galaxies' rotations without introducing the invisible "dark matter" while general relativity needs the existence of the mysterious "dark energy" to be able to predict the ongoing cosmic accelerating expansion. On one side, it is important to seek for descriptions for these dark contents as ingredients of physics beyond the standard model. On the other side, it is also reasonable to think that general relativity may not be accurate enough to describe the larger systems. The latter idea has brought the field of modified gravity to life.

One of the remarkable models of modified gravity comes from a very simple attempt to extend the gravity theory. In the context of particle physics, gravitational force is carried by a force carrier known as a graviton. The graviton that carries the force in general relativity actually has no mass. In this sense, one can think of a simple generalization of the gravity theory by introducing nonzero mass to the graviton. This idea is the essense of the massive gravity theories. Massive gravity started its story in 1939 when Fierz and Pauli proposed a linear theory of massive gravity [1]. Though it is actually a direct generalization from the linearized general relativity, it was proved by van Dam, Veltman, and Zakharov that the Fierz-Pauli theory cannot recover the linearized general relativity [2, 3]. To cure this problem, Vainshtein proposed that this problem can be lifted if nonlinearities are included into the theory [4]. Furthermore, the nonlinearities required in this aspect must be chosen very carefully, pointed out by Boulware and Deser, to avoid the ghost instability, and instability where one of the degrees of freedom has negative kinetic energy [5]. Embracing those requirements, de Rham, Gabadadze, and Tolley proposed a nonlinear theory of massive gravity in 2010, dubbed dRGT massive gravity [6, 7]. This nonlinear theory has given opportunities to cosmologists to study our universe in the context of massive gravity. In particular, this massive gravity theory can describe the cosmic accelerating expansion in terms of the nonvanishing graviton mass.

Though the dRGT massive gravity provides a possible answer to the cosmic acceleration, it faces a serious problem. In particular, the corresponding cosmological solution is unstable in the context of the dRGT massive gravity [8]. This means that we cannot trust the cosmological solution obtained from the theory. To fix this problem, there are a lot of attempts to modify the dRGT massive gravity. One of those attempts is to let the graviton mass to be a varying function of a scalar field, dubbed mass-varying massive gravity [9, 10, 11, 12, 13]. This model, however, implies that the graviton mass, as a varying function, keeps shrinking its mass as the universe expands. With hopes of fixing this problem, in this work we allow the graviton mass to vary due to not only a scalar field but also its kinetic term. Moreover, the scalar field is now governed by the k-essence lagrangian. We then investigate this proposed model in terms of its cosmology. We found that such an allowance can cause modifications on the gravity sector as if there effectively exists the dark matter. As a result, not only that the graviton mass contributes as the effective dark energy but also it gives rise to the effective dark matter. Furthermore, since this model can provide descriptions of both dark energy and dark matter in terms of varying graviton mass, we also investigate this model in the context of the cosmic coincidence problem. Particularly, the cosmic coincidence problem is a "coincidence" of the universe to have dark matter as much as dark energy. In the aspects of the well-known ΛCDM model, dark energy and dark matter vary in a very drastic way over cosmological time such that the fact that the universe constituted of dark matter as much as dark energy, as observed nowadays, appears as a coincidence. We then attempted to explain this coincidence in the context of mass-varying massive gravity. In this work, we review the key contents of this mass-varying massive gravity presented in Ref. [13].

2. The model and the equations of motion

In this section we explicitly expressed the model action and discuss about important features through the equations of motion. Following the previously mentioned concepts, one can construct the corresponding action as follows [13],

$$S = \int d^4x\sqrt{-g}\left[\frac{M_p^2}{2}R[g] + V(X,\phi)(\mathcal{L}_2[g,f] + \alpha_3\mathcal{L}_3[g,f] + \alpha_4\mathcal{L}_4[g,f]) + P(X,\phi)\right], \tag{1}$$

where R is a Ricci scalar corresponding to the physical metric $g_{\mu\nu}$, $V(X,\phi)$ corresponds to the varying graviton mass which is a function of the scalar field ϕ and its kinetic term $X \equiv -\frac{1}{2}g^{\mu\nu}\nabla_\mu\phi\nabla_\nu\phi$, $P(X,\phi)$ is the k-essence lagrangian governing the scalar field ϕ, and the

IOP Conf. Series: Journal of Physics: Conf. Series **883** (2017) 012005 doi:10.1088/1742-6596/883/1/012005

terms denoted by $\mathcal{L}_i$'s are expressed as follows,

$$\mathcal{L}_2[g,f] = \frac{1}{2}\left([\mathcal{K}]^2 - [\mathcal{K}^2]\right), \tag{2}$$

$$\mathcal{L}_3[g,f] = \frac{1}{3!}\left([\mathcal{K}]^3 - 3[\mathcal{K}][\mathcal{K}^2] + 2[\mathcal{K}^3]\right), \tag{3}$$

$$\mathcal{L}_4[g,f] = \frac{1}{4!}\left([\mathcal{K}]^4 - 6[\mathcal{K}]^2[\mathcal{K}^2] + 3[\mathcal{K}^2]^2 + 8[\mathcal{K}][\mathcal{K}^3] - 6[\mathcal{K}^4]\right), \tag{4}$$

where the square brackets $[\cdot]$ denote the trace of an object inside them and we have defined a building-block tensor $\mathcal{K}^\mu_{\ \nu}$ as

$$\mathcal{K}^\mu_{\ \nu} = \delta^\mu_{\ \nu} - \left(\sqrt{g^{-1}f}\right)^\mu_{\ \nu}. \tag{5}$$

Here $f_{\mu\nu}$ is a reference metric which allows us to construct the nontrivial and nonlinear massive gravity. Moreover, the introduction of this reference metric $f_{\mu\nu}$ helps in the sense that the general covariance can now be equipped into the massive gravity via the following expression,

$$f_{\mu\nu} = \partial_\mu \varphi^\rho \partial_\nu \varphi^\sigma \tilde{f}_{\rho\sigma}, \tag{6}$$

where φ^μ's are known as the Stückelberg fields which, since they transform as scalars, then implement the general covariance into the reference metric. For simplicity, we can choose the form of $\tilde{f}_{\mu\nu}$ to be of the same form as the physical metric in order to ease our calculation.

Since we focus on cosmological implications, we then consider this model with the physical metric of the Friedmann-Lemaître-Robertson-Walker (FLRW) form. In particular, we use the following ansatz for the physical metric,

$$ds^2 = g_{\mu\nu}dx^\mu dx^\nu = -N(t)^2 dt^2 + a(t)^2 \Omega_{ij}(x) dx^i dx^j, \tag{7}$$

$$\Omega_{ij}(\varphi) = \delta_{ij} + \frac{k\delta_{ia}\delta_{jb}\varphi^a\varphi^b}{1 - k\delta_{lm}\varphi^l\varphi^m}, \tag{8}$$

where N is a lapse function that allows the time reparameterization, a is a scale factor denoting the overall scale of the 3-dimensional space, Ω_{ij} is the 3-dimensional spatial metric which classifies the spatial geometry to be open, flat, or closed one. For the reference metric, it is chosen to be of the same form as the FLRW metric as

$$\tilde{f}_{\mu\nu}d\varphi^\mu d\varphi^\nu = -n(\varphi^0)^2\left(d\varphi^0\right)^2 + \alpha(\varphi^0)^2\Omega_{ij}(\varphi)d\varphi^i d\varphi^j, \tag{9}$$

where $n(\varphi^0), \alpha(\varphi^0)$ are a lapse function and a scale factor in the reference sector.

From the action in Eq. (1) we can find the equations of motion, assuming the unitary gauge $\varphi^\mu = x^\mu$ and the spatially flat geometry $k = 0$, as follows,

$$M_p^2\left(3H^2 + 3\frac{k}{a^2}\right) = -3VF + \rho_X - P, \tag{10}$$

$$M_p^2\left(\frac{2\dot{H}}{N} + 3H^2 + \frac{k}{a^2}\right) = -3VF + VF_{,\bar{X}}\left(\bar{X} - \eta\right) - P, \tag{11}$$

$$\frac{\dot{V}}{V} = NH\left(1 - h\bar{X}\right)\frac{F_{,\bar{X}}}{G}, \tag{12}$$

$$Na^3\left(3V_{,\phi}\left(F - G\eta\right) + P_{,\phi}\right) = \frac{d}{dt}\left[\frac{a^3}{\sqrt{2X}}\rho_X\right], \tag{13}$$

IOP Conf. Series: Journal of Physics: Conf. Series **883** (2017) 012005 doi:10.1088/1742-6596/883/1/012005

where the dot represents the time derivative and we have used the following definitions,

$$\begin{aligned}
H &\equiv \frac{\dot{a}}{aN}, \qquad h \equiv \frac{H_\alpha}{H}, \qquad H_\alpha \equiv \frac{\dot{\alpha}}{\alpha n}, \\
\eta &\equiv \frac{n}{N}, \qquad \rho_X \equiv 6XV_{,X}\left(F - G\eta\right) + 2XP_{,X}, \\
F &\equiv \left(2 + \frac{4}{3}\alpha_3 + \frac{1}{3}\alpha_4\right) - \left(3 + 3\alpha_3 + \alpha_4\right)\bar{X} + \left(1 + 2\alpha_3 + \alpha_4\right)\bar{X}^2 - \left(\alpha_3 + \alpha_4\right)\frac{\bar{X}^3}{3}, \qquad (14) \\
G &\equiv \frac{1}{3}\left(3 + 3\alpha_3 + \alpha_4\right) - \left(1 + 2\alpha_3 + \alpha_4\right)\bar{X} + \left(\alpha_3 + \alpha_4\right)\bar{X}^2 - \alpha_4\frac{\bar{X}^3}{3}, \\
\bar{X} &\equiv \frac{\alpha}{a}.
\end{aligned}$$

Note that since ρ_X lies on the right-hand side of the Friedmann equation in Eq. (10), we can view ρ_X as an energy density of an effective matter. For simplicity, we may assume that both the graviton mass function and the k-essence lagrangian are functions of the kinetic term X only,

$$V(X,\phi) = V(X), \qquad P(X,\phi) = P(X). \qquad (15)$$

From this assumption, Eq. (13) can be rewritten as

$$\frac{d}{dt}\rho_X + 3HN\rho_X = \frac{\dot{X}}{2X}\rho_X. \qquad (16)$$

This equation can be thought of as a continuity equation of a nonrelativistic matter of energy density ρ_X which can flow to other sectors due to the nonzero term on the right-hand side. In other words, this effective matter corresponding to ρ_X is not conserved due to the flow. Consequently, we can imply that this model of mass-varying massive gravity, with some specific assumptions, can give rise to effective dust which from now on will be thought of as effective dark matter of energy density ρ_X.

Since we have seen that this model can give rise to the effective dark matter, it is important to investigate whether this model can provide a description of the dark energy responsible to the cosmic accelerating expansion or not. To this end, we may perform a simple analysis by first defining the following definitions,

$$\rho_g \equiv -3VF + 6XV_{,X}\left(F - G\eta\right), \qquad p_g \equiv 3VF - VF_{,\bar{X}}\left(\bar{X} - \eta\right). \qquad (17)$$

Note that ρ_g can be viewed as contributions arising from the graviton mass alone which can be seen from the Friedmann equation in Eq. 10 that ρ_g is the term involving the graviton mass function V on the right-hand side of the equation, and p_g is the corresponding pressure. In the following calculation we will investigate whether the matter described by the energy density ρ_g and the pressure p_g can be a candidate for the dark energy. We assume that the equation of state parameter $w_g \equiv \frac{p_g}{\rho_g}$ of this matter satisfies $w_g = -1$. By equating w_g to -1 we assume that this matter content shares the same properties as the cosmological constant in ΛCDM model (see Ref. [14] for a review). As a consequence, we obtain the following condition,

$$6XV_{,X}\left(F - G\eta\right) = VF_{,\bar{X}}\left(\bar{X} - \eta\right). \qquad (18)$$

In a very simple case where $F, G, \bar{X}, \eta, F_{,\bar{X}}$ are constants, we obtain a very simple condition for V as

$$V \propto X^{\lambda/2}, \qquad \lambda = \text{constant}. \qquad (19)$$

This form of V indicates that the system in consideration is in the dark energy phase where the cosmic expansion is driven as if there is the cosmological constant. This encourages us to utilize this form of V in the following calculations.

At this point we have seen roughly that in this model of massive gravity the gravity can be modified as if there is the effective dust in the system. Moreover, the model can also describe the cosmic accelerating expansion, at least in the cosmological constant manner. These features encourage us to investigate in the next section in terms of the cosmic coincidence problem. This cosmic coincidence problem is a problem arising from the fact that we observed possible amounts of dark energy and dark matter in our universe. In particular, we found that there exists as much dark energy as there is dark matter in our universe (those amounts are in the same order of magnitude). This is somehow unusual in the context of ΛCDM model since each matter content varies so drastically over cosmological time in the ΛCDM universe (see Ref. [15]). Thus, the fact that the universe contains dark matter and dark energy, each of the same order of magnitude, seems to be a "coincidence" in the ΛCDM context. Since this mass-varying massive gravity model has a possibility of being able to explain the coexistence between dark matter and dark energy, it is worthwhile to consider the model in the aspect of the cosmic coincidence problem.

3. Dynamical Analysis & Cosmic Coincidence Problem

In order to investigate the model in the aspect of the cosmic coincidence problem, since the equations of motion we found in Eq. (10) - Eq. (13) are cumbersome, we can use the methods of dynamical analysis to extract some useful information out of those equations. We first use the specific form of V we obtained earlier and we also assume the form of P as follows,

$$V(X) = V_0 X^{\lambda/2}, \qquad P(X) = P_0 X^{\gamma/2}, \tag{20}$$

where λ, γ are constants. We then define density parameters for each matter content in the right-hand side of the Friedmann equation in Eq. (10) as follows,

$$x \equiv -\frac{FV}{M_p^2 H^2}, \qquad y \equiv \frac{\rho_X}{3M_p^2 H^2}, \qquad z \equiv -\frac{P}{3M_p^2 H^2}. \tag{21}$$

These density parameters are governed by the following dynamical equations,

$$x' = 3x\left(y + sx - \frac{s}{r}\right), \tag{22}$$

$$y' = 3y\left(y + sx - 1 - \frac{s}{\lambda r}\right), \tag{23}$$

$$1 = x + y + z, \tag{24}$$

$$y = -\lambda x(1-r) - z\gamma, \tag{25}$$

where the prime denotes the derivative with respect to $\log a$ and

$$r \equiv \frac{G\eta}{F}, \qquad s \equiv \frac{F_{,\bar{X}}(\bar{X} - \eta)}{3F}. \tag{26}$$

We can also find an effective equation of state parameter for the entire system to be

$$w_{eff} = -1 + y + xs. \tag{27}$$

From these dynamical equations one can find fixed points of the system at which the configuration of the system will stay as it is or, in other words, the system will be stationary if its configuration corresponds to any of these points. All of the fixed points are shown in Table 1.

Table 1. Summary of the properties of the fixed points.

Name	x	y	z	w_{eff}	existence	stability
(a)	0	0	1	-1	$\gamma = 0$	$\frac{s}{r} \geq 0$
(b)	$\frac{1}{r}$	0	$1 - \frac{1}{r}$	$-1 + \frac{s}{r}$	$\gamma = \lambda$	$\frac{\lambda}{1-\lambda} \leq \frac{s}{r} < 0$
(c)	0	$1 + \frac{s}{\lambda r}$	$-\frac{s}{\lambda r}$	$\frac{s}{\lambda r}$	$\gamma = 1 + \frac{\lambda r}{s}$	$\frac{\lambda}{1-\lambda} < \frac{s}{r} < -1$
(d)	$\frac{1}{1+\lambda(r-1)}$	$\frac{\lambda(r-1)}{1+\lambda(r-1)}$	0	$\frac{1}{\lambda-1}$	$\lambda = \frac{s}{s-r}$	$0 < \lambda < 1$
(e)	$\frac{1+(\lambda-1)z_0}{1+\lambda(r-1)}$	$-\frac{\lambda(1-r(z_0+1))}{1+\lambda(r-1)}$	z_0	$\frac{1}{\lambda-1}$	$\lambda = \gamma = \frac{s}{s-r}$	$0 < \lambda < 1$

The fixed point (a) and (b) correspond to the situations found in ΛCDM model and the dRGT massive gravity respectively. The point (c) can be a great candidate for the dust domination period since the configuration for y can be nonzero. If this is the case, then we should have a very large value of λ for the effective equation of state parameter to approach zero. For the cosmic acceleration period, it is reasonable to think that one of the point (d) or (e) can represent this period since w_{eff} can be close to -1 when $\lambda \to 0$. If these points represent dynamics of the universe which evolves from the dust dominated era to the dark energy/dark matter era, despite being constant, λ must be very large during the dust domination while it must be very small at the time of cosmic acceleration. This suggests that λ must change in time or be a varying function. By promoting λ to be a function, λ is governed by the following dynamical equation,

$$\lambda' = \frac{6s}{r}\left(\frac{\lambda}{2} - (1+\Gamma)\right), \tag{28}$$

where $\Gamma \equiv XV_{,XX}/V_{,X}$. We then compute numerical solutions of the evolutions of each matter content, including that of radiation, in order to verify the idea that the system can evolve from the dust dominated era to the dark energy/dark matter era. To obtain the numerical solutions, we first set an initial condition at $\log a = 0$ corresponding to the (present time) so that ratio of the amount of dark energy to that of dark matter at that time is $7:3$, which is chosen for simplicity to represent the observed amounts of the dark energy and the dark matter. Note that the amount of radiation is also included which is set to be of order 10^{-5} to ensure that there is only small portion of radiation at the present time. Moreover, we also choose appropriate values of λ and s and let $\lambda = \gamma$ at $\log a = 0$ to ensure that the configuration at the present time corresponds to either point (d) or point (e) and the equaion of state parameter is close to -1. Furthermore, we choose Γ to be a small negative number so that λ grows slowly as the time ($\log a$) goes backward. This parameter setup is equivalent to setting specific values to all of the model parameters, including α_3 and α_4. According to their characteristics, the solutions are classified into three following kinds,

(i) **Massive Gravity Domination** This solution is explicitly shown in Figure 1 where the system evolves without the presence of k-essence lagrangian, in other words $z = 0$ all the time. Unfortunately, this solution only tells us that without the presence of nonzero z we cannot have a universe filled with both dark energy and dark matter at the same time as the universe will eventually evolve towards a full domination of dark energy (blue-dashed line, representing x) where no dark matter is presented (red-dotted line, representing y).

(ii) **Late-time Universe 1** This solution, shown in Figure 2, shows us the possibility to solve the cosmic coincidence problem. In other words, the universe at late-time is filled with comparable amounts between dark energy (black-solid line, representing $x + z$) and

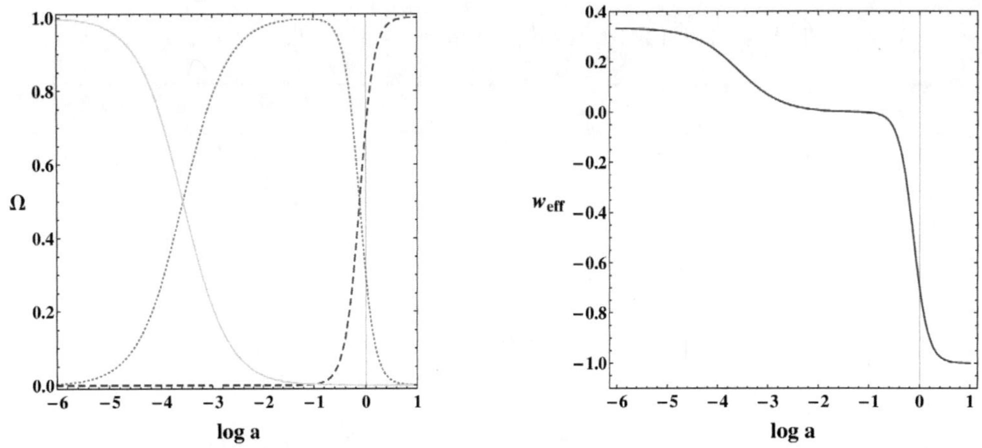

(a) (b)

Figure 1. The plot (a) shows evolutions of various kinds of matter and energy. The green-solid line corresponds to the evolution of radiation, the red-dotted line corresponds to the evolution of the effective dark matter (y), and the blue-dashed line corresponds to that of the effective dark energy (x). The plot (b) shows the corresponding effective equation of state parameter throughout the evolution. Note that at late time the system is filled mostly with the effective dark energy.

dark matter (red-dotted line, representing y). Note that in this case the dark energy is represented by the quantity $x+z$ rather than that in the previous case which is represented only by x (because $z = 0$). The effective equation of state parameter at the dark energy phase is significantly less than -1.

(iii) **Late-time Universe 2** This solution is just another possibility to solve the cosmic coincidence problem. From Figure 3, the entire evolution is restricted by the condition $\gamma = \lambda$. The result of such a condition is that it allows the universe filled with comparable amounts of dark energy (black-solid line, representing $x + z$) and dark matter (red-dotted line, representing y). However, this solution requires a fine-tuning of the initial conditions in the numerical computations. This poses another problem into this solution which is known as the fine-tuning problem.

4. Discussion

We have seen various features of this model of mass-varying massive gravity where the graviton mass is promoted to be a function of a scalar field and also its kinetic term. Such an allowance, as far as its cosmology is concerned, can modify the gravity in this theory as if there is the effective dark matter in the system of interest. We then investigate in a very simple way a possibility of this model, while the effective dark matter is presented, to enter the phase of the cosmic accelerating expansion and we obtain the following suitable form of the graviton mass function $V(X) \propto X^{\lambda/2}$. Such a structure is investigated whether it can provide a meaningful cosmology or not via the method of dynamical analysis. Through it we found five fixed points, one corresponds to the ΛCDM cosmology, one is similar to the cosmology in dRGT massive gravity, one can be thought of as the dust dominated period, and other two have a chance to solve the cosmic coincidence problem since they can provide the cosmic acceleration while the effective dark matter is presented. The points of our interest are the dust dominated point and the cosmic accelerating expansion points. If these points are to explain each phase of the

IOP Conf. Series: Journal of Physics: Conf. Series **883** (2017) 012005 doi:10.1088/1742-6596/883/1/012005

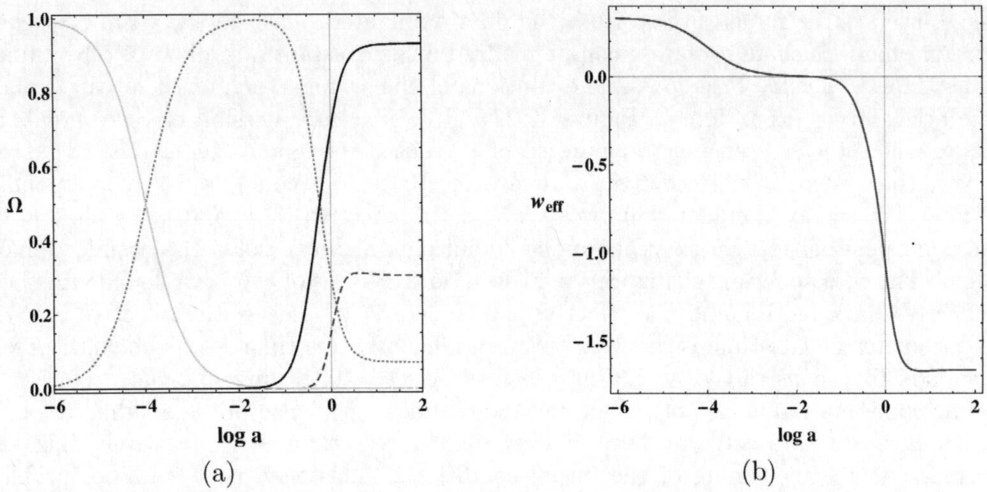

Figure 2. The plot (a) shows evolutions of various kinds of matter and energy. The green-solid line corresponds to the evolution of radiation, the red-dotted line corresponds to the evolution of the effective dark matter (y), the blue-dashed line corresponds to the quantity x, and the black-solid line corresponds to the "total" dark energy (which is $x + z$). The plot (b) shows the corresponding effective equation of state parameter throughout the evolution. Note that at late time the system is filled with both effective dark matter and effective dark energy, accompanying with the effective equation of state parameter being significantly less than -1.

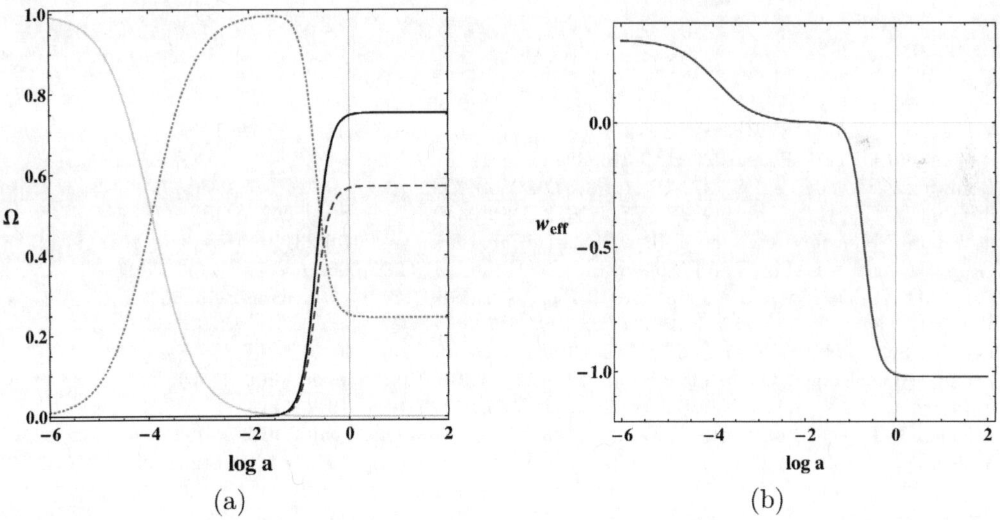

Figure 3. The plot (a) shows evolutions of various kinds of matter and energy. The green-solid line corresponds to the evolution of radiation, the red-dotted line corresponds to the evolution of the effective dark matter (y), the blue-dashed line corresponds to the quantity x, and the black-solid line corresponds to the "total" dark energy (which is $x + z$). The right plot shows the corresponding effective equation of state parameter throughout the evolution where the condition $\gamma = \lambda$ is satisfied. Note that at late-time the system is filled with both effective dark matter and effective dark energy, accompanying with the effective equation of state parameter being slightly less than -1.

universe, the system should evolve from the dust dominated phase to the cosmic accelerating expansion phase. Such an evolution can be realized if the constant λ is allowed to be a function. We then show explicitly three possible evolutions of the system, each of them can explain the cosmic accelerating expansion, in Figure 1, 2, and 3. The first possibility corresponds to the situation where the k-essence lagrangian is not presented thoughout the evolution, i.e. $z = 0$. However, the system of this configuration evolves towards a configuration corresponding to the fully dark energy dominated universe where the effective dark matter is almost absent. The second possibility is an example of the evolution which can solve the cosmic coincidence problem. The corresponding configuration at late time consists of comparable amounts between the effective dark matter and the effective dark energy. Moreover, the effective equation of state parameter at late-time is required to be significantly less than -1. The third possibility corresponds to a constrained evolution which obeys $\gamma = \lambda$ throughout the evolution. The late-time configuration is capable of solving the cosmic coincidence problem while the effective equation of state parameter at late time is slightly less than -1. However, this kind of evolution requires fine-tuning of the initial conditions. This then posts another problem on the configuration which is known as the fine-tuning problem. From our previous analysis, this model of mass-varying massive gravity can alleviate the cosmic coincidence problem, though in some aspects a new problem occurs, like the fine-tuning problem.

Acknowledgments
This work is a conference paper based on a work in Ref. [13]. Moreover, this work was supported by the National Research Foundation of Korea(NRF) grant funded by the Korea government(MSIP) (No.2016R1C1B1010107).

References
[1] Fierz M and Pauli W 1939 *Proc. Roy. Soc. Lond.* **A173** 211–232
[2] van Dam H and Veltman M J G 1970 *Nucl. Phys.* **B22** 397–411
[3] Zakharov V I 1970 *JETP Lett.* **12** 312 [Pisma Zh. Eksp. Teor. Fiz.12,447(1970)]
[4] Vainshtein A I 1972 *Phys. Lett.* **B39** 393–394
[5] Boulware D G and Deser S 1972 *Phys. Rev.* **D6** 3368–3382
[6] de Rham C and Gabadadze G 2010 *Phys. Rev.* **D82** 044020 (*Preprint* 1007.0443)
[7] de Rham C, Gabadadze G and Tolley A J 2011 *Phys. Rev. Lett.* **106** 231101 (*Preprint* 1011.1232)
[8] Gumrukcuoglu A E, Lin C and Mukohyama S 2012 *JCAP* **1203** 006 (*Preprint* 1111.4107)
[9] Huang Q G, Piao Y S and Zhou S Y 2012 *Phys. Rev.* **D86** 124014 (*Preprint* 1206.5678)
[10] Wu D J, Piao Y S and Cai Y F 2013 *Phys. Lett.* **B721** 7–12 (*Preprint* 1301.4326)
[11] Leon G, Saavedra J and Saridakis E N 2013 *Class. Quant. Grav.* **30** 135001 (*Preprint* 1301.7419)
[12] Huang Q G, Zhang K C and Zhou S Y 2013 *JCAP* **1308** 050 (*Preprint* 1306.4740)
[13] Tannukij L and Wongjun P 2016 *Eur. Phys. J.* **C76** 17 (*Preprint* 1511.02164)
[14] Frieman J, Turner M and Huterer D 2008 *Ann. Rev. Astron. Astrophys.* **46** 385–432 (*Preprint* 0803.0982)
[15] Velten H E S, vom Marttens R F and Zimdahl W 2014 *Eur. Phys. J.* **C74** 3160 (*Preprint* 1410.2509)

Constraints on dark matter annihilation by Planck

Chalit Muanglay, Maneenate Wechakama and Brandon K. Cantlay

Department of Physics, Faculty of Science, Kasetsart University, Bangkok, Thailand

E-mail: chalit.ml@gmail.com, mwechakama@gmail.com

Abstract. We investigate the production of electrons and positrons in the Milky Way within the context of dark matter annihilation. Upper limits on the relevant cross-section are obtained by Planck data at different wavelengths with recent measurements of the positron spectra in the solar neighbourhood by AMS02. We consider synchrotron emission in the microwave bands. According to our results, the dark matter annihilation cross-section into electron-positron pairs should not be higher than the canonical value for a thermal relic if the mass of the dark matter candidate is smaller than a few GeV.

1. Introduction

Dark matter can be indirectly detected through the signatures of standard model particles produced by its annihilation or decay (see e.g. [1]). The recent results from indirect detection experiments in the solar neighbourhood have suggested the possibility that such a signature has been seen. In particular, the PAMELA experiment has pointed a significant excess of electrons and positrons above the expected smooth astrophysical background [2]. If these results are interpreted in terms of dark matter annihilation, then an abundant population of high-energy $e^{\pm}$ is being created everywhere in the Galactic dark matter halo, with the associated synchrotron emission in the Galactic magnetic field.

The present work focuses on the astrophysical signatures of dark matter annihilation into electron-positron pairs. We try to impose constraints on the relevant cross-section by comparing the predictions of an analytic model of particle propagation with a set of observational data obtained from Planck. In addition to the photon data, we also consider the recent measurements of the local positron spectra performed by AMS02 [3].

Rather than focusing on a particular dark matter candidate, we adopt a model-independent approach (see e.g. [4]), in which all the injected particles are created with the same initial energy E_0, of the order of the mass of the dark matter particle. Since this mass is usually much larger than the rest mass of the electron, electrons and positrons will be relativistic at the moment of their creation. However, they can efficiently lose their energy through different processes, such as inverse Compton scattering (ICS), synchrotron radiation, Coulomb collisions, bremsstrahlung, and ionization. We will often use the Lorentz factor γ to express the energy $E = \gamma m_{\mathrm{e}} c^2$ of the annihilation products, where m_{e} denotes the rest mass of electron, and c is the speed of light.

2. Model predictions

2.1. Electron-positron propagation

As in our previous work [4], the propagation of electrons and positrons through the interstellar medium (ISM) is determined by the diffusion-loss equation

$$\begin{aligned}\frac{\partial}{\partial t}\frac{\mathrm{d}n}{\mathrm{d}\gamma}(\boldsymbol{x},\gamma) &= \nabla\left[K(\boldsymbol{x},\gamma)\nabla\frac{\mathrm{d}n}{\mathrm{d}\gamma}(\boldsymbol{x},\gamma)\right] \\ &+ \frac{\partial}{\partial\gamma}\left[b(\boldsymbol{x},\gamma)\frac{\mathrm{d}n}{\mathrm{d}\gamma}(\boldsymbol{x},\gamma)\right] \\ &+ Q(\boldsymbol{x},\gamma). \end{aligned} \tag{1}$$

We assume a diffusion coefficient of the form

$$K(\gamma) = K_0\gamma^{\delta}. \tag{2}$$

The energy loss rate

$$b(\boldsymbol{x},\gamma) \equiv -\frac{\mathrm{d}\gamma}{\mathrm{d}t}(\boldsymbol{x},\gamma) = \sum_i b_i(\boldsymbol{x},\gamma) \tag{3}$$

is a sum over the relevant physical processes, and the source term $Q(\boldsymbol{x},\gamma)$ represents the instantaneous electron-positron injection rate.

Given enough time, the electron-positron population will approach a steady state distribution, $\frac{\partial}{\partial t}\frac{\mathrm{d}n}{\mathrm{d}\gamma}(\boldsymbol{x},\gamma) = 0$. Assuming that $b(\boldsymbol{x},\gamma)$ varies smoothly in space, the particle spectrum fulfills the relation

$$\frac{\partial y(\boldsymbol{x},\gamma)}{\partial\gamma} + \frac{K(\gamma)}{b(\gamma)}\nabla^2 y(\boldsymbol{x},\gamma) = -Q(\boldsymbol{x},\gamma), \tag{4}$$

where

$$y(\boldsymbol{x},\gamma) \equiv b(\gamma)\frac{\mathrm{d}n}{\mathrm{d}\gamma}(\boldsymbol{x},\gamma). \tag{5}$$

Imposing $\frac{\mathrm{d}n}{\mathrm{d}\gamma}(\boldsymbol{x},\gamma) = 0$ at infinity, one obtains the Green's function

$$G(\boldsymbol{x},\gamma,\boldsymbol{x}_{\mathrm{s}},\gamma_{\mathrm{s}}) = \frac{\exp\left(-\frac{|\boldsymbol{x}-\boldsymbol{x}_{\mathrm{s}}|^2}{2\Delta\lambda^2}\right)}{(2\pi\Delta\lambda^2)^{3/2}}\,\Theta(\gamma-\gamma_{\mathrm{s}}) \tag{6}$$

and either the image charges method or an expansion over the eigenfunctions of the linear differential operator may be used to derive the Green's function for other boundary conditions (see e.g. [5]). The electron-positron spectrum is thus given by

$$\frac{\mathrm{d}n}{\mathrm{d}\gamma}(\mathbf{x},\gamma) = \frac{1}{b(\mathbf{x},\gamma)}\int_{\gamma}^{\infty}\mathrm{d}\gamma_{\mathrm{s}}\int_0^{\infty}\mathrm{d}^3\mathbf{x}_{\mathrm{s}}\frac{\exp\left(-\frac{|\mathbf{x}-\mathbf{x}_{\mathrm{s}}|^2}{2\Delta\lambda^2}\right)}{(2\pi\Delta\lambda^2)^{3/2}}Q(\mathbf{x}_{\mathrm{s}},\gamma_{\mathrm{s}}), \tag{7}$$

where the quantity

$$\Delta\lambda^2 = \lambda^2(\gamma) - \lambda^2(\gamma_{\mathrm{s}}) \tag{8}$$

is related to the characteristic diffusion length of the electrons and positrons, γ_{s} denotes their initial energy, and the variable λ is defined as

$$\lambda^2(\gamma) = \int_{\gamma}^{\infty}\frac{2K(\gamma)}{b(\gamma)}\mathrm{d}\gamma. \tag{9}$$

Considering the dark matter halo as a spherically-symmetric source, the spatial integral can be reduced to one dimension, and the electron-positron spectrum is finally given by the expression

$$\frac{\mathrm{d}n}{\mathrm{d}\gamma}(r,\gamma) = \frac{1}{b(\gamma)} \frac{\exp\left(-\frac{r^2}{2\Delta\lambda^2}\right)}{(2\pi r^2 \Delta\lambda^2)^{1/2}} \times \Bigg\{ \int_\gamma^\infty \mathrm{d}\gamma_\mathrm{s} \int_0^\infty \mathrm{d}r_\mathrm{s}\, r_\mathrm{s} \exp\left(-\frac{r_\mathrm{s}^2}{2\Delta\lambda^2}\right) \times \left[\exp\left(\frac{r r_\mathrm{s}}{\Delta\lambda^2}\right) - \exp\left(-\frac{r r_\mathrm{s}}{\Delta\lambda^2}\right)\right] Q(r_\mathrm{s},\gamma_s) \Bigg\}. \tag{10}$$

2.2. Loss rates

Electrons and positrons can lose their energy by several physical processes as they move through the ISM. We consider ICS of cosmic microwave background (CMB), starlight and infrared photons, synchrotron radiation, Coulomb collisions, bremsstrahlung, and ionization of neutral hydrogen atoms.

The energy loss rates depend on the energy of the particle. High-energy electrons and positrons mainly lose energy by ICS, e.g. [6]. We compute the total power radiated by a single electron using the formalism, based on the Klein-Nishina cross-section. In the non-relativistic regime, the loss function can be approximated as

$$b_\mathrm{ICS}(\gamma) = \frac{4}{3}\frac{\sigma_\mathrm{T}}{m_\mathrm{e}c}\gamma^2 U_\mathrm{rad}, \tag{11}$$

where σ_T is the Thomson cross-section. The combined radiation energy density of the CMB, starlight (SL), and infrared (IR) light from thermal dust emission (see e.g. [7], [8]) is represented by three grey bodies,

$$U_\mathrm{rad} = \frac{4\,\sigma_\mathrm{SB}}{c}\left(T_\mathrm{CMB}^4 + \mathcal{N}_\mathrm{SL} T_\mathrm{SL}^4 + \mathcal{N}_\mathrm{IR} T_\mathrm{IR}^4 \right) \tag{12}$$

where T_i and $\mathcal{N}_i$ represent the effective temperature and the normalization of each component, respectively, and σ_SB is the Stefan-Boltzmann constant.

Synchrotron radiation is another important loss mechanism at high energies. The expression for the loss rate is similar to that of non-relativistic ICS, substituting the radiation energy density in equation (11) by the magnetic energy density, $U_\mathrm{B} = B^2/(8\pi)$, where B is the intensity of the magnetic field:

$$b_\mathrm{syn}(\gamma) = \frac{4}{3}\frac{\sigma_\mathrm{T}}{m_\mathrm{e}c}\gamma^2 U_\mathrm{B}. \tag{13}$$

For lower-energy electrons and positrons, Coulomb interactions with the thermal plasma must be taken into account. The loss rate is approximately [9]

$$b_\mathrm{Coul}(\gamma) \approx 1.2 \times 10^{-12} n_\mathrm{e} \left[1 + \frac{\ln(\gamma/n_\mathrm{e})}{75}\right] \ \mathrm{s}^{-1}, \tag{14}$$

where n_e is the number density of thermal electrons.

Collisions with thermal ions and electrons also produce radiation through bremsstrahlung. The loss rate due to bremsstrahlung can be approximated as [10]

$$b_\mathrm{brem}(\gamma) \approx 1.51 \times 10^{-16} n_\mathrm{e}\gamma \left[\ln(\gamma) + 0.36\right] \ \mathrm{s}^{-1}. \tag{15}$$

IOP Conf. Series: Journal of Physics: Conf. Series **883** (2017) 012006 doi:10.1088/1742-6596/883/1/012006

Additional energy losses come from the ionization of hydrogen atoms. The loss rate is given in [11],

$$b_{\rm ion}(\gamma) = \frac{q_{\rm e}^4 n_{\rm H}}{8\pi\epsilon_0^2 m_{\rm e}^2 c^3 \sqrt{1-\frac{1}{\gamma^2}}} \times \left[\ln \frac{\gamma(\gamma^2-1)}{2\left(\frac{I}{m_e c^2}\right)^2} - \left(\frac{2}{\gamma} - \frac{1}{\gamma^2}\right)\ln 2 + \frac{1}{\gamma^2} + \frac{1}{8}\left(1-\frac{1}{\gamma}\right)^2 \right], \tag{16}$$

where $n_{\rm H}$ is the number density of hydrogen atoms, $q_{\rm e}$ is the electron charge, ϵ_0 is the permittivity of free space, and I is the ionization energy of the hydrogen atom. The number density of thermal electrons and neutral atoms can be expressed in terms of the total ISM gas density $\rho_{\rm g}$ and the ionization fraction $X_{\rm ion}$ as

$$n_e = \frac{\rho_{\rm g}}{m_{\rm p}} X_{\rm ion} \tag{17}$$

and

$$n_{\rm H} = \frac{\rho_{\rm g}}{m_{\rm p}}(1 - X_{\rm ion}), \tag{18}$$

respectively.

2.3. Source term

Since the electrons and positrons in our model originate from the annihilation of dark matter particles, the instantaneous production rate at any given point can be expressed as

$$Q(r,\gamma) = \eta\ n_{\rm dm}(r)\ n_{\rm dm^*}(r)\ \langle\sigma v\rangle_{\rm e^\pm} \frac{{\rm d}N_{\rm e^\pm}}{{\rm d}\gamma}(\gamma), \tag{19}$$

where $n_{\rm dm}$ and $n_{\rm dm^*}$ denote the number densities of dark matter particles and anti-particles, respectively, $\langle\sigma v\rangle_{\rm e^\pm}$ is the thermal average of the annihilation cross-section times the dark matter relative velocity, and $\frac{{\rm d}N_{\rm e^\pm}}{{\rm d}\gamma}$ is the injection spectrum of electrons and positrons in the final state. For self-conjugate dark matter particles, $n_{\rm dm} = n_{\rm dm^*} = \frac{\rho_{\rm dm}}{m_{\rm dm}}$ and $\eta = 1/2$ in order to avoid double counting; else, $n_{\rm dm} = n_{\rm dm^*} = \frac{1}{2}\frac{\rho_{\rm dm}}{m_{\rm dm}}$ and $\eta = 1$.

We consider self-conjugate dark matter particles throughout this work and assume that each annihilation event injects one electron and one positron with roughly the same energy $\gamma_0 \sim m_{\rm dm}/m_{\rm e}$,

$$\frac{{\rm d}N_{\rm e^\pm}}{{\rm d}\gamma}(\gamma) = 2\ \delta(\gamma - \gamma_0), \tag{20}$$

where $\delta(\gamma-\gamma_0)$ denotes a Dirac delta function. Although this is a rather coarse approximation, it has the advantage of being model-independent. For self-conjugate dark matter particles, we obtain

$$Q(r,\gamma) = \left[\frac{\rho_{\rm dm}(r)}{m_{\rm dm}}\right]^2 \langle\sigma v\rangle_{\rm e^\pm}\ \delta(\gamma-\gamma_0). \tag{21}$$

We consider a spherically-symmetric halo, described by a density profile of the form

$$\rho_{\rm dm}(r) = \frac{\rho_{\rm s}}{\left(\frac{r}{r_{\rm s}}\right)^\alpha \left(1+\frac{r}{r_{\rm s}}\right)^{3-\alpha}}, \tag{22}$$

where $r_{\rm s}$ and $\rho_{\rm s}$ denote a characteristic density and radius of the halo, respectively, and α is the inner logarithmic slope of the density profile. Local inhomogeneities that would boost the expected signal, such as small-scale clumpiness or the presence of subhaloes, are not taken into account.

2.4. Surface brightness profile

Once the electron-positron spectrum is computed, the emission coefficient for photons of frequency ν is given by the integral

$$j_\nu(r,\nu) = \frac{1}{4\pi}\int_1^\infty \frac{\mathrm{d}n}{\mathrm{d}\gamma}(r,\gamma)\; l(\gamma,\nu)\; \mathrm{d}\gamma \tag{23}$$

of the electron-positron spectrum $\frac{\mathrm{d}n}{\mathrm{d}\gamma}(r,\gamma)$ times the specific luminosity $l(\gamma,\nu)$ emitted at frequency ν by a single electron or positron with Lorentz factor γ. The intensity from any given direction in the sky is simply the integral along the line of sight of the emission coefficient. Since we assume a spherically-symmetric source and boundary conditions, it will only depend on the angular separation θ with respect to the Galactic centre,

$$I_\nu(\theta,\nu) = \int_0^\infty j_\nu(r,\nu)\; \mathrm{d}s, \tag{24}$$

where s represents the distance along the line of sight, and the radial distance r to the centre of the Milky Way at any point along the ray is

$$r = \sqrt{x^2 + y^2}, \tag{25}$$

with $x = s\sin\theta$, $y = s\cos\theta - R_\odot$, and $R_\odot = 8.5$ kpc (the distance of the Sun from the Galactic centre).

The contribution of synchrotron radiation, which dominates at low photon energies, can be estimated as [6]

$$l_{\rm syn}(\gamma,\nu) = \frac{\sqrt{3}\,q_{\rm e}^3 B}{m_{\rm e}c^2}\; R[\chi(\gamma)], \tag{26}$$

where $m_{\rm e}$ and $q_{\rm e}$ denote the electron mass and charge, respectively, B is the intensity of the magnetic field, and the function $R(\chi)$ is defined as [12]

$$R(\chi) \equiv 2\chi^2 \left[K_{\frac{4}{3}}(\chi) K_{\frac{1}{3}}(\chi) - \frac{3}{5}\chi \left\{ K_{\frac{4}{3}}^2(\chi) - K_{\frac{1}{3}}^2(\chi) \right\} \right]. \tag{27}$$

In this expression, K refers to the modified Bessel function, and the normalized frequency

$$\chi \equiv \frac{\nu}{3\gamma^2 \nu_c} \tag{28}$$

is expressed in terms of the cyclotron frequency

$$\nu_c \equiv \frac{q_{\rm e} B}{2\pi m_{\rm e} c}. \tag{29}$$

2.5. Astrophysical parameters

The photon intensity from the synchrotron depends on the astrophysical parameters that determine the propagation and energy losses of the relativistic particles. We calculate the electron-positron spectrum as described in expression (10), and then estimate the photon intensity according to expression (24).

Our canonical model assumes a dark matter density profile with $\alpha = 1$ [13], $r_{\rm s} = 17$ kpc and $\rho_{\rm s}c^2 = 0.35$ GeV cm^{-3}, consistent with dynamical models of the Milky Way. The virial mass of the Galaxy is thus 10^{12} M$_\odot$, and the local dark matter density is $\rho_{\rm dm}(r_\odot)\, c^2 = 0.3$ GeV cm^{-3}. The ISM is mainly composed of neutral hydrogen atoms ($X_{\rm ion} = 0$) with number density

Model	K_0 [kpc^2 s^{-1}]	δ
MED	1.76×10^{-18}	0.70

Table 1. The model of the diffusion coefficient, following the parameterization $K(\gamma) = K_0\gamma^\delta$. The model MED has been proposed by [14].

Model	$\mathscr{N}_{\rm SL}$ $T_{\rm SL} = 3481$ K	$\mathscr{N}_{\rm IR}$ $T_{\rm IR} = 40.6$ K
ISRF (I)	2.7×10^{-12}	7.0×10^{-5}
canonical	1.7×10^{-11}	7.0×10^{-5}

Table 2. Normalization of the grey-body models describing the interstellar radiation field, adopted from [15]. In our canonical model, we use the values appropriate for the Galactic centre in order to compute the ICS and synchrotron emission. For the electron-positron spectrum at the Solar neighbourhood, we use ISRF(I).

$\rho_{\rm g}/m_{\rm p} \sim 1$ cm^{-3}, and it is permeated by a tangled magnetic field whose intensity is $B \sim 6$ μG throughout the Galaxy.

For the diffusion coefficient (see equation 2), we consider the MED model discussed by [14], summarized in Table 1. We will also use a model of the ISRF (adopted from [15]) where the photon intensity is represented by grey-body components. The normalizations and effective temperatures of the light emitted by the Galactic stars and dust are quoted in Table 2.

3. Observational data

In order to constrain the production of relativistic electrons and positrons in the Milky Way, we consider observations of the whole sky at different wavelengths by using data from 9 channels of Planck observation at microwave wavelengths, which dominated by synchrotron emission. We take the full-resolution coadded temperature maps for each of the 9 frequency bands (30 GHz, 44 GHz, 70 GHz, 100 GHz, 143 GHz 217 GHz, 353 GHz, 545 GHz and 857 GHz) [16].

Since we are interested in a spherically-symmetric component, we may follow a simple, conservative procedure in order to mask the emission from the Galactic disk and individual point sources without relying on any particular foreground model. For each frequency, we compute the average intensity $I(\theta)$ in 180 bins as a function of the angular separation θ from the Galactic centre. We also estimate the standard deviation $\sigma(\theta)$ within each bin, as well as the average standard deviation

$$\sigma_{\rm ave} = \frac{\sum_{i=1}^{n} \sigma(\theta_i)}{n}, \tag{30}$$

where $n = 180$ is the total number of the bins. We then start an iterative procedure, where all pixels more than $3\,\sigma_{\rm ave}$ away from $I(\theta)$ are discarded until convergence is achieved.

The average intensity $I(\theta)$ for each wavelength are shown in Figures 1. Besides these observational data, we also consider the positrons spectrum in the solar neighbourhood by AMS02.

4. Constraints on the dark matter cross-section

Once the emission from the Galactic disc and the most prominent point sources is excluded, the remaining spherically-averaged component can be used to place upper limits on the cross-section for dark matter annihilation into electron-positron pairs.

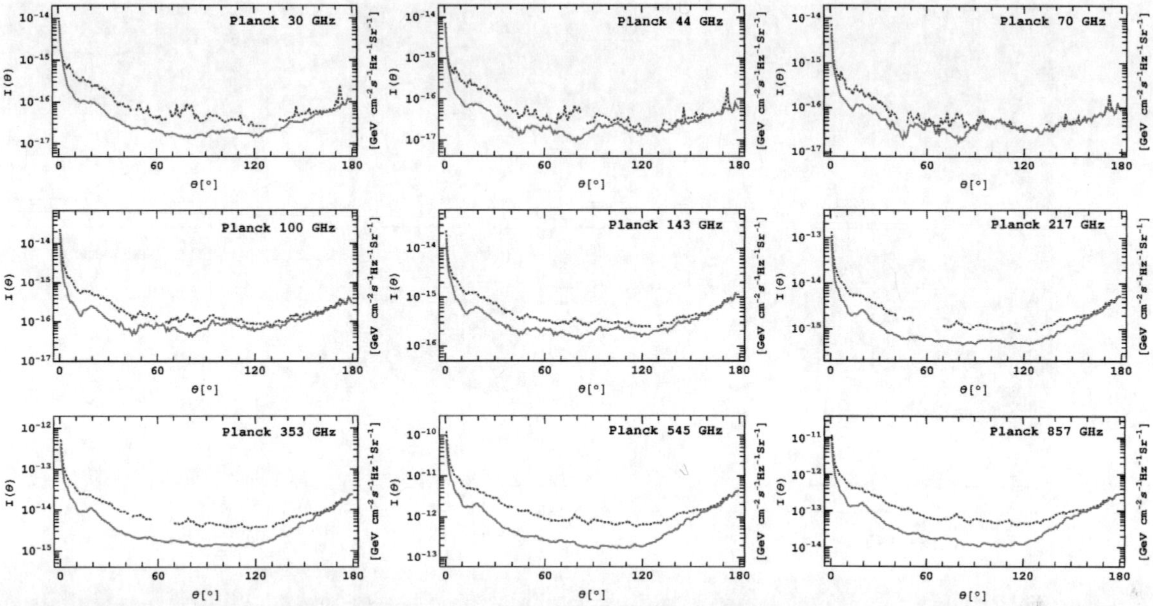

Figure 1. Planck spherically-averaged intensities. Solid red lines represent the original mean intensity $I_0(\theta)$, while dotted blue lines correspond to the final intensity $I(\theta)$ after discarding the outliers .

First of all, model intensities are computed according to the scheme described in Section 2. We consider the injection energy (i.e. the mass of the dark matter particle) as a free parameter and investigate values of the initial Lorentz factor γ_0 between 2×10^3 and 2×10^7.

For synchrotron radiation, particle diffusion makes the intensity profile shallower, especially at high injection energies. In general, one can say that photons of a given frequency trace electrons and positrons within a certain energy range. If that range is close to E_0, these particles would have just been injected, and therefore the effects of particle propagation should be small, whereas, away from E_0, these electrons and positrons would have traveled a significant distance from the point of injection, and the surface brightness profile will become considerably shallower.

In addition to the photons arriving from the centre of the Milky Way, the dark matter annihilation cross-section $\langle \sigma v \rangle_{e^\pm}$ is also strongly constrained by the observed abundance of relativistic electrons and positrons in the solar neighbourhood. In particular, we consider the recent measurements of the *positron* spectrum by AMS02.

Our constraints are derived by imposing that the predicted amount of electrons and/or positrons does not exceed the observed values for *any* Lorentz factor γ. Given the energy dependence of the observed spectrum, $\left[\frac{\mathrm{d}n}{\mathrm{d}E}\right]_{\rm obs} \sim E^{-3}$, and the energy losses, $b(E) \sim E^2$, the most restrictive constraint comes from the spectrum near the injection energy, where propagation can be safely neglected and $\left[\frac{\mathrm{d}n}{\mathrm{d}E}\right]_{\rm model} \approx \frac{Q_0}{b} \propto E^{-2}$. The maximum production rate allowed by the data can then be expressed as

$$Q_0(r_\odot) < b(\gamma_0) \left[\frac{\mathrm{d}n}{\mathrm{d}E}\right]_{\rm obs}(\gamma_0), \tag{31}$$

and one arrives to the condition

$$\langle \sigma v \rangle_{e^\pm}(\gamma_0) < \left[\frac{m_{\rm dm}}{\rho_{\rm dm}(r_\odot)}\right]^2 b(\gamma_0) \left[\frac{\mathrm{d}n}{\mathrm{d}E}\right]_{\rm obs}(\gamma_0) \tag{32}$$

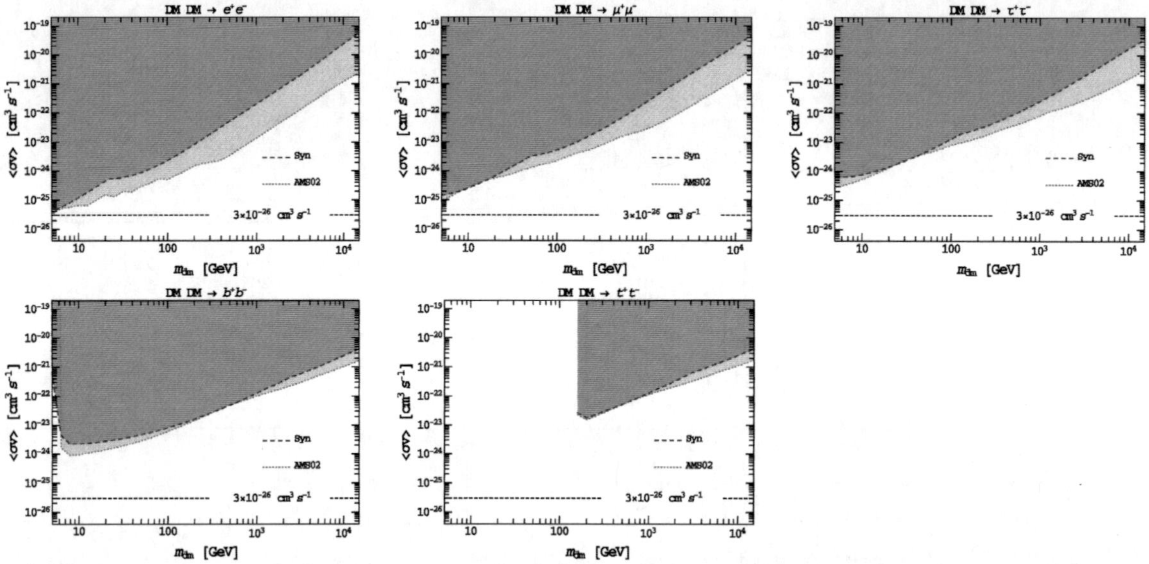

Figure 2. Upper limits on the dark matter annihilation cross-section for different annihilation channels derived by comparing the predicted synchrotron (blue dashed line)and the positron spectrum are shown by the red areas. The horizontal dotted line indicates the value $\langle \sigma v \rangle_{e^\pm} = 3 \times 10^{-26}$ cm s^{-1}.

in order not to overproduce the observed signal.

From the point of view of particle physics, dark matter annihilation directly into electron-positron pairs is arguably not the most natural channnel. In most models, dark matter annihilates into heavier products, and then these particles produce lower-energy electrons and positrons as secondaries. More precisely, we consider different source functions, replacing the Dirac delta in equation (20) by the appropriate injection spectrum. We used the electron-positron fluxes at production computed by [17], including electroweak corrections, for all the leptonic channels, as well as for annihilation into top and bottom quarks.

The upper limits on the annihilation cross-section into each channel are plotted in Figure 2. As can be readily seen in the figure, the tightest constraints are provided by the positron spectrum in the solar neighbourhood and synchrotron emission limit the production cross-section at lower energies. The constraints obtained for dark matter annihilation into muon-antimuon pairs are somewhat weaker than electron-positron pairs (by a factor of a few, especially at the interesting regime of low dark matter masses) due to the softer injection spectrum.

For annihilation into τ particles, the upper limits imposed synchrotron emission and the local positron spectrum are even less stringent. Similar conclusions may be reached for annihilation into top or bottom quarks.

5. Summary and conclusions

We have investigated the constraints on the dark matter annihilation cross-section into electron-positron pairs by comparing the predictions of an analytic model of particle propagation with a set of observational data obtained from Planck. We have compared the expected emission from synchrotron radiation within the Milky Way with 9 maps of the sky at different frequencies (30 GHz, 44 GHz, 70 GHz, 100 GHz, 143 GHz 217 GHz, 353 GHz, 545 GHz and 857 GHz). A straightforward statistical criterion has been followed in order to mask the most obvious astrophysical signals (i.e. the emission from the Galactic disc and prominent point sources), and

IOP Conf. Series: Journal of Physics: Conf. Series **883** (2017) 012006 doi:10.1088/1742-6596/883/1/012006

observational upper limits are derived from the remaining spherically-symmetric component. In addition, we have also imposed that the predicted abundance of positrons in the solar neighbourhood does not exceed the measurements by AMS02.

According to our results, the dark matter annihilation cross-section into electron-positron pairs should not be higher than the canonical value for a thermal relic if the mass of the dark matter candidate is smaller than a few GeV. The upper limit on the cross-section is set by the local positron spectrum for low values of the injection energy.

Acknowledgments
This research is supported in part by the Graduate Program Scholarship from the Graduate School, Kasetsart University.

References
[1] Bertone G 2010 *Particle Dark Matter* (Cambridge: Cambridge University Press)
[2] Adriani O, et al., PAMELA Collaboration 2009 *Nature.* **458** 607
[3] Aguilar M, et al. 2014 *Physical Review Letters.* **113** 121102
[4] Wechakama M, and Ascasibar Y 2014 *MNRAS* **439** 566
[5] Baltz E A and Edsjö J 1999 *Phys. Rev. D* **59** 023511
[6] Sarazin C L 1999 *ApJ* **520** 529
[7] Porter T A and Strong A W 2005 *International Cosmic Ray Conference* **4** 77
[8] Porter T A, et al. 2008 *ApJ* **682** 400
[9] Rephaeli Y 1979 *ApJ* **227** 364
[10] Blumenthal G R and Gould R J 1970 *Review of Modern Physics* **42** 237
[11] Longair M S 1981 *High energy astrophysics: An Informal Introduction for Students of Physics and Astronomy* (Cambridge: Cambridge University Press)
[12] Ghisellini G, Guilbert P W, and Svensson R 1988 *ApJL* **334** 5
[13] Navarro J F, Frenk C S, and White S D M 1996 *ApJ* **462** 563
[14] Donato F, et al. 2004 *Phy. Rew. D.* **69** 063501
[15] Cirelli M, and Panci P 2009 *Nuclear Physics B* **821** 399
[16] Planck Collaboration 2016 *Astronomy & Astrophysics* **594** A1
[17] Cirelli M et al. 2011 *JCAP* **3** 51

IOP Conf. Series: Journal of Physics: Conf. Series **883** (2017) 012007 doi:10.1088/1742-6596/883/1/012007

Slow-roll inflation and the BB-mode correlation spectrum of the Cosmic Microwave Background

N Malsawmtluangi[1] and P K Suresh[2]
School of Physics, University of Hyderabad
P.O. Central University, Hyderabad 500046, India.
E-mail: [1]tei.naulak@uohyd.ac.in; [2]pkssp@uohyd.ernet.in

Abstract. The BB-mode correlation angular spectrum of cosmic microwave background is studied for several inflation models. A comparative study of the resulting angular power spectrum with the recent joint analysis data of the BICEP2/Keck Array and Planck shows constraint to several slow-roll inflation models of cosmology.

1. Introduction

The theory of cosmic inflation is the most widely known scenario proposed to resolve several issues associated with the standard model of cosmology, like large scale isotropy and homogeneity, horizon and flatness problems, monopole problem, etc.,. A large number of inflation models have been studied over several years [1, 2, 3, 4, 5]. It is believed that quantum fluctuations during inflation gave rise to primordial fluctuations like scalar and tensor perturbations. The scalar perturbations seeded the formation of large scale structures in the universe and the tensor perturbations represent the primordial gravitational waves (GWs).

The primordial GWs were generated during inflation by the strong and variable gravitational field of the early universe through the mechanism of parametric amplification of the zero-point quantum fluctuations which transforms the initial vacuum state with no particle into a multi-particle quantum state [6, 7] called squeezed vacuum state [8, 9]. All primordial GW modes start in the same vacuum state and this vacuum state gets transformed into a strongly squeezed state. Due to the squeezing effect, the variance of the GW mode's phase is strongly squeezed while the variance of its amplitude is being enhanced so that the uncertainty product remains intact. The parameter of squeezing grows all the way up in the amplifying regime and can vary from zero in the vacuum state up to a very large value by the end of the amplifying regime. From the vacuum state to the inflationary period, it has grown from zero up to 1 ($0 \leq r_s \leq 1$) which we use for the present study. The primordial GWs are believed to exist in a strongly squeezed vacuum state.

The primordial GWs are believed to have left an imprint on the cosmic microwave background (CMB) anisotropy in the form of B-mode polarization. Hence, if the primordial GWs do exist in the squeezed vacuum state, then the effect of squeezing should also be observed in the B-mode angular power spectrum of CMB [10]. This phenomenon is employed as a platform to provide further constraint on the inflation models. The results are compared with the recent BICEP2/Keck Array and Planck collaboration data [11] which sets the upper and lower bounds for the tensor-to-scalar ratio respectively as $r < 0.07$ and $r \simeq \mathcal{O}(10^{-3})$ [12].

IOP Conf. Series: Journal of Physics: Conf. Series **883** (2017) 012007 doi:10.1088/1742-6596/883/1/012007

2. Primordial GWs in the Squeezed Vacuum State

The perturbed metric for a flat Friedmann-Lemaitre-Robertson-Walker (FLRW) universe can be written as,

$$dS^2 = a^2(\eta)[-d\eta^2 + (\delta_{ij} + h_{ij})dx^i dx^j], \tag{1}$$

where a is the scale factor and η is the conformal time defined by $d\eta = \frac{dt}{a}$, h_{ij} is a transverse-traceless perturbation of space-time, $|h_{ij}| \ll \delta_{ij}$,

$$\partial_i h^{ij} = 0, \quad \delta^{ij} h_{ij} = 0.$$

where δ_{ij} is the flat space metric.

The GW field $h_{ij}(\mathbf{x}, \eta)$ can be decomposed into a collection of Fourier modes as,

$$h_{ij}(\mathbf{x}, \eta) = \frac{C}{(2\pi)^{\frac{3}{2}}} \int_{-\infty}^{+\infty} \frac{d^3\mathbf{k}}{\sqrt{2k}} \sum_{p=1}^{2} [h_k^{(p)}(\eta) c_k^{(p)} \varepsilon_{ij}^{(p)}(\mathbf{k}) e^{i\mathbf{k}.\mathbf{x}} + h_k^{(p)*}(\eta) c_k^{(p)\dagger} \varepsilon_{ij}^{(p)*}(\mathbf{k}) e^{-i\mathbf{k}.\mathbf{x}}], \tag{2}$$

where $C = \sqrt{16\pi} l_{pl}$ is the quantum normalization of the GW field and $l_{pl} = \sqrt{G}$ is Planck's length.

The creation operator $c_k^{\dagger}$ and the annihilation operator c_k satisfy the relationships $\left[c_k^{(p)}, c_{k'}^{(p')\dagger}\right] = \delta_{pp'} \delta^3(k - k')$, $\left[c_k^{(p)}, c_{k'}^{(p')}\right] = \left[c_k^{(p)\dagger}, c_{k'}^{(p')\dagger}\right] = 0$. The Heisenberg equations of motion govern the evolution of these operators.

The two polarization states $\varepsilon_{ij}^{(p)}$, $p = 1, 2$ are symmetric and transverse-traceless and satisfy the conditions $\varepsilon_{ij}^{(p)} \delta^{ij} = 0$, $\varepsilon_{ij}^{(p)} k^i = 0$, $\varepsilon_{ij}^{(p)} \varepsilon^{(p')ij} = 2\delta_{pp'}$, $\varepsilon_{ij}^{(p)}(\text{-}\mathbf{k}) = \varepsilon_{ij}^{(p)}(\mathbf{k})$. These polarizations are linear and are called the plus (+) polarization and cross (×) polarization. The contribution from each polarization is same. Hence, from here onward, the superscript (p) is dropped for convenience.

The GW mode $h_k(\eta)$ can be rescaled in terms of the mode function μ_k as,

$$h_k = \frac{\mu_k}{a}. \tag{3}$$

The mode function can have the following form,

$$\mu_k(\eta) = u_k(\eta) + v_k^*(\eta), \tag{4}$$

where $u_k(\eta)$ and $v_k(\eta)$ are complex functions.

The mode function satisfies the equation of motion,

$$\mu_k'' + \left(k^2 - \frac{a''}{a}\right) \mu_k = 0, \tag{5}$$

where prime indicates the derivative with respect to the conformal time η.

The two-point correlation function of the modes leads to the power spectrum of tensor perturbations as,

$$\langle h_k h_{k'}^* \rangle = \frac{2\pi^2}{k^3} P_T(k) \delta^3(\mathbf{k} - \mathbf{k}'), \tag{6}$$

where P_T is the gravitational wave power spectrum.

The squeezed vacuum state is defined as [13],

$$|\xi\rangle = S(\xi)|0\rangle, \tag{7}$$

where $S(\xi)$ is the squeezing operator which can be written as,

$$S(\xi) = \exp\left[\frac{1}{2}\xi^* b^2 - \frac{1}{2}\xi b^{\dagger 2}\right], \tag{8}$$

where $\xi = r_s e^{i\gamma}$ is a complex number, r_s is the squeezing parameter and γ is the squeezing angle. The unitary transformations of the squeezing operator S on the annihilation and creation operators lead to

$$\begin{aligned} S^\dagger(\xi) b S(\xi) &= b\cosh r_s - b^\dagger e^{i\gamma}\sinh r_s, \\ S^\dagger(\xi) b^\dagger S(\xi) &= b^\dagger \cosh r_s - b e^{-i\gamma}\sinh r_s. \end{aligned} \tag{9}$$

Using Eqs.(2) and (9), the two point correlation function in the squeezed vacuum state can be written as,

$$\langle h_k h_{k'}^* \rangle = \frac{C^2}{a^2}\left[(1 + 2\sinh^2 r_s)|\mu_k|^2 + \frac{1}{2}\sinh 2r_s(\mu_k^2 e^{i\gamma} + \mu_k^{*2} e^{-i\gamma})\right]\delta^3(\mathbf{k} - \mathbf{k}'). \tag{10}$$

From Eq.(6) and Eq.(10), we get the power spectrum in the squeezed vacuum state as,

$$P_T(k) = \frac{k^3}{2\pi^2}\frac{C^2}{a^2}\left[(1 + 2\sinh^2 r_s)|\mu_k|^2 + \frac{1}{2}\sinh 2r_s(\mu_k^2 e^{i\gamma} + \mu_k^{*2} e^{-i\gamma})\right]. \tag{11}$$

In the quasi de Sitter universe during inflation, one can define the relation between the conformal time and the scale factor as $a(\eta) = \frac{-1}{H\eta(1-\epsilon)}$, where H is the Hubble parameter, the slow roll parameter is $\epsilon = \frac{m_{pl}^2}{2}\left(\frac{V'}{V}\right)^2$ and V is the potential of the scalar field. For small ϵ, $\vartheta = \frac{3}{2} + \epsilon$, and $n_T = -2\epsilon = 3 - 2\vartheta$.

For constant ϵ, the equation of motion can be written as [14],

$$\mu_k'' + \left[k^2 - \frac{1}{\eta^2}\left(\vartheta^2 - \frac{1}{4}\right)\right]\mu_k = 0. \tag{12}$$

The general solution for the above equation (Eq.(12)) is,

$$\mu_k(\eta) = \sqrt{-\eta}[C_1(k)\mathcal{H}_\vartheta^{(1)}(-k\eta) + C_2(k)\mathcal{H}_\vartheta^{(2)}(-k\eta)], \tag{13}$$

where $\mathcal{H}_\vartheta^{(1)}$ and $\mathcal{H}_\vartheta^{(2)}$ are the Hankel functions of the first and second kind, and C_1 and C_2 are the constants of integration. Using the flat spacetime solutions $\mu_k^0(\eta) = \frac{1}{\sqrt{2k}}e^{-ik\eta}$, the constants of integration become $C_1(k) = \frac{\sqrt{\pi}}{2}\exp\left[i\left(\vartheta + \frac{1}{2}\right)\left(\frac{\pi}{2}\right)\right]$ and $C_2(k) = 0$.

Then Eq.(13) implies that for long wavelength limit ($k << aH$),

$$\mu_k(\eta) = e^{i(\vartheta - \frac{1}{2})(\frac{\pi}{2})} 2^{\vartheta - \frac{3}{2}} \frac{\Gamma(\vartheta)}{\Gamma\left(\frac{3}{2}\right)} \frac{1}{\sqrt{2k}} (-k\eta)^{\frac{1}{2} - \vartheta}. \tag{14}$$

Using Eq.(14) in Eq.(11), the power spectrum in the superhorizon limit ($k << aH$) is,

$$P_T(k) = C^2\left(\frac{H}{2\pi}\right)^2\left(\frac{k}{aH}\right)^{3-2\vartheta}\left[1 + 2\sinh^2 r_s + \sinh 2r_s\cos\left(\gamma + \left(\vartheta - \frac{1}{2}\right)\pi\right)\right]. \tag{15}$$

Finally, taking $A_T(k_0) = C^2\left(\frac{H_{k0}}{2\pi}\right)^2$ as the normalization constant for the tensor spectrum, where H_{k0} is the Hubble parameter at $aH = k_0$, k_0 being the pivot frequency, we get the power spectrum for the gravitational waves in terms of the tensor spectral index n_T as,

$$P_T(k) = A_T(k_0)\left(\frac{k}{k_0}\right)^{n_T}\left[1 + 2\sinh^2 r_s + \sinh 2r_s\cos\left(\gamma + (2 - n_T)\frac{\pi}{2}\right)\right]. \tag{16}$$

Eq.(16) represents the power spectrum for gravitational waves in the squeezed vacuum state.

IOP Conf. Series: Journal of Physics: Conf. Series **883** (2017) 012007 doi:10.1088/1742-6596/883/1/012007

3. Inflationary scenario

In the simplest inflationary scenario a canonical scalar field ϕ called the inflaton drives the accelerated expansion of the early universe.

The equation of motion for the inflaton is given by,

$$\ddot{\phi} + 3H\dot{\phi} + V'(\phi) = 0, \tag{17}$$

where dot and prime indicate derivatives with respect to time (t) and the field (ϕ) respectively. V is the effective potential of the inflaton. The Hubble parameter H gives the strength of friction and is determined by the energy density of the scalar field,

$$\rho_\phi = \frac{\dot{\phi}^2}{2} + V,$$

so that the Friedmann equation can be written as,

$$H^2 = \frac{1}{3m_{pl}^2}\left(\frac{1}{2}\dot{\phi}^2 + V(\phi)\right). \tag{18}$$

In the slow-roll limit, the energy density of the inflaton is dominated by its potential energy, $\frac{\dot{\phi}^2}{2} \ll V$. Hence in this limit, called the slow-roll limit, the Hubble parameter and the inflaton potential are related as,

$$H^2 \simeq \frac{V}{3m_{pl}^2}. \tag{19}$$

This condition is characterized by the slow-roll parameters which can be defined in terms of the inflaton potential (V) and its derivatives as,

$$\begin{aligned} \epsilon &\equiv \frac{m_{pl}^2}{2}\left(\frac{V'}{V}\right)^2, \\ \eta &\equiv m_{pl}^2\left(\frac{V''}{V}\right), \end{aligned} \tag{20}$$

and so on. Inflation lasts as long as the slow-roll conditions are satisfied, i.e., $\epsilon \ll 1$ and $|\eta| \ll 1$. Once these slow roll conditions are violated, inflation ends followed by reheating and decay of the inflaton which is followed by particle production, which is the case for most inflation models.

In the slow-roll approximation, the power spectra of the scalar perturbation (P_S) and the tensor perturbation (P_T) generated outside the horizon can be given in terms of the potential by,

$$P_S \simeq \frac{1}{12\pi^2 m_{pl}^6}\frac{V^3}{V'^2}\bigg|_{k=aH}, \tag{21}$$

$$P_T \simeq \frac{1}{3\pi^2 m_{pl}^4}V\bigg|_{k=aH}, \tag{22}$$

where $k = aH$ indicates that both H and V are evaluated at the time when the mode with wave number k crosses the horizon. For all calculations, we take the scalar power spectrum to be $P_S = 2.43 \times 10^{-9}$.

The tensor-to-scalar ratio can be written in terms of the parameter ϵ as,

$$r \equiv \frac{P_T(k)}{P_S(k)} \simeq 16\epsilon. \tag{23}$$

This parameter measures the strength of the tensor spectrum relative to that of the scalar spectrum and can be very useful to distinguish between the many inflation models.

IOP Conf. Series: Journal of Physics: Conf. Series **883** (2017) 012007 doi:10.1088/1742-6596/883/1/012007

4. Inflation Models

In this section, we briefly discuss several slow-roll inflation models for which the upper bound for the tensor-to-scalar ratio is set to be $r < 0.07$ and lower limit is $r \simeq \mathcal{O}(10^{-3})$ and are constrained with various CMB observations [12]. We compute the slow roll parameters, tensor spectral index, tensor power spectrum and tensor-to-scalar ratio for the following slow-roll inflation models.

4.1. Natural Inflation model

In this model the inflaton is considered to be the Nambu Goldstone boson which arises whenever global symmetry is broken [15, 16].

The effective potential for the model is given by,

$$V(\phi) = M^4\left[1 + \cos\left(\frac{\phi}{f}\right)\right], \tag{24}$$

where $f/m_{pl} = 10^2$ is the energy scale at which symmetry is broken, $M/m_{pl} = 10^{-2}$.

Using Eq.(20), the slow-roll parameters are calculated as,

$$\epsilon = 1.29 \times 10^{-3}, \tag{25}$$

$$\eta = 1.24 \times 10^{-3}. \tag{26}$$

The equation of state parameters are calculated as $r = 2.06 \times 10^{-2}$ and $n_T = -2.58 \times 10^{-3}$. The tensor power spectrum is calculated to be $P_T = 5.027 \times 10^{-11}$.

4.2. Arctan Inflation model

This model is often studied as a toy model [17, 18]. In the present scenario, we consider a large field model which starts at a large ϕ which later evolves to a minimum at the origin.

The effective potential for the model is,

$$V(\phi) = M^4\left[1 - \arctan\left(\frac{\phi}{\mu}\right)\right], \tag{27}$$

where $\mu/m_{pl} = 10^{-2}$ is a free parameter which characterizes the typical vacuum expectation value at which inflation takes place, $M/m_{pl} = 10^{-3}$.

Using Eq.(20), the slow-roll parameters are found to be,

$$\epsilon = 8.62 \times 10^{-4}, \tag{28}$$

$$\eta = 3.0 \times 10^{-2}. \tag{29}$$

The equation of state parameters are calculated as $r = 1.38 \times 10^{-2}$ and $n_T = -1.72 \times 10^{-3}$. The tensor power spectrum is obtained as, $P_T = 3.35 \times 10^{-11}$.

4.3. Inverse Monomial Inflation Model

This model is considered here in the context of quintessential inflation [19, 20, 21]; the inflaton need not necessarily decay and reheating arises naturally even when the potential does not have a global minimum, radiation is created via gravitational particle production.

The effective potential for the model is,

$$V(\phi) = M^4\left(\frac{\phi}{m_{pl}}\right)^{-p}, \tag{30}$$

here $p = 3$ is a positive parameter, $M/m_{pl} = 10^{-1}$.

Using Eq.(20), the slow-roll parameters are calculated as,

$$\epsilon = 1.25 \times 10^{-4}, \quad (31)$$
$$\eta = 3.33 \times 10^{-4}. \quad (32)$$

The corresponding equation of state parameters are calculated as $r = 2.0 \times 10^{-3}$ and $n_T = -2.50 \times 10^{-4}$. The tensor power spectrum for this model is calculated as $P_T = 4.86 \times 10^{-12}$.

4.4. Hybrid Inflation model

This is a multi-scalar field [22, 23] model where ϕ field drives the inflation and the inflation ends abruptly by the symmetry breaking of σ field.

The effective potential of hybrid inflation model is,

$$V = \frac{1}{4\lambda}(M^2 - \lambda\sigma^2)^2 + \frac{1}{2}m^2\phi^2 + \frac{1}{2}g^2\phi^2\sigma^2, \quad (33)$$

where $M = 1.21 \times 10^{16}$ GeV, $m = 3.65 \times 10^{11}$ GeV, $\lambda = 1$, $g = 8 \times 10^{-4}$.

Using Eq.(20), the slow-roll parameters are found to be,

$$\epsilon = 2.65 \times 10^{-4},$$
$$\eta = 1.47 \times 10^{-4}. \quad (34)$$

The equation of state parameters for this model are calculated as $r = 4.24 \times 10^{-3}$ and $n_T = -5.3 \times 10^{-4}$. The tensor power spectrum for this model is obtained as $P_T = 1.03 \times 10^{-11}$.

4.5. Loop Inflation Model

In this model, the flatness of the inflaton potential is altered by symmetry breaking along the flatness of the potential which produces quantum radiative corrections in which one loop order correction takes the form of a logarithmic function [24, 25, 26].

The effective potential for this model can be written as,

$$V(\phi) = M^4\left[1 + \alpha\ln\left(\frac{\phi}{m_{pl}}\right)\right], \quad (35)$$

where $\alpha = g^2/16\pi^2$ tunes the strength of radiative effects, $M = 10^{16}$ GeV.

Using Eq.(20), the slow-roll parameters are found to be,

$$\epsilon = 3.09 \times 10^{-3}, \quad (36)$$
$$\eta = -2.06 \times 10^{-2}. \quad (37)$$

The corresponding equation of state parameters are $r = 4.34 \times 10^{-2}$ and $n_T = -6.18 \times 10^{-3}$. The tensor power spectrum is calculated as, $P_T = 1.2 \times 10^{-10}$.

4.6. Coleman-Weinberg Inflation model

In this scenario, the potential is introduced in the context of spontaneous symmetry breaking generated by radiative corrections [27, 28, 29].

The effective potential for the model is,

$$V(\phi) = M^4\left[1 + \alpha\left(\frac{\phi}{\sigma}\right)^4\ln\left(\frac{\phi}{\sigma}\right)\right], \quad (38)$$

IOP Conf. Series: Journal of Physics: Conf. Series **883** (2017) 012007 doi:10.1088/1742-6596/883/1/012007

where $\alpha = 4e$, $M = 10^{16}$ GeV, $\sigma = 10m_{pl}$ sets the typical vacuum expectation value at which inflation takes place.

Using Eq.(20), the slow-roll parameters are found to be,

$$\epsilon = 4.86 \times 10^{-4}, \quad (39)$$

$$\eta = -4.42 \times 10^{-2}. \quad (40)$$

The equation of state parameters are obtained as $r = 7.77 \times 10^{-3}$ and $n_T = -9.72 \times 10^{-4}$. The tensor power spectrum is calculated as, $P_T = 1.89 \times 10^{-11}$.

4.7. Quadratic Chaotic Inflation model with radiative corrections

This model is a simple quadratic chaotic inflation model [30] studied under the assumption that the scalar field interacts with the fermion field thus leading to the quantum radiative correction which takes the form of a logarithmic function [31]. Introduction of radiative corrections lowers the tensor-to-scalar ratio and the power level of the model.

The effective potential with radiative correction is,

$$V(\phi) = \frac{1}{2}m^2\phi^2 - \frac{g^4}{16\pi^2}\phi^4 \ln\left(\frac{\phi}{m_{pl}}\right), \quad (41)$$

where g represents the Yukawa coupling, $m = 3.44 \times 10^{12}$ GeV.

Using Eq.(20), the slow-roll parameters are found to be,

$$\epsilon = 1.86 \times 10^{-3},$$

$$\eta = 1.86 \times 10^{-3}. \quad (42)$$

The equation of state parameters are obtained as $r = 2.98 \times 10^{-2}$ and $n_T = -3.72 \times 10^{-3}$. The tensor power spectrum for this model is calculated as, $P_T = 7.25 \times 10^{-11}$.

5. BB-mode correlation spectrum

It is believed that the scalar and tensor fluctuations have left their own signatures on the CMB and can be realized in the form of E-mode and B-mode polarizations respectively [32, 33, 34]. The BB-mode correlation angular power spectrum of CMB is given by [35, 36],

$$C_l^{BB} = (4\pi)^2 \int dk k^2 P_T(k) \left| \int_0^{\eta_0} d\eta g(\eta) h_k(\eta) \times \left\{ (8x + 2x^2\partial_x)\frac{j_l(x)}{x^2} \right\}_{x=k(\eta_0-\eta)} \right|^2, \quad (43)$$

where $g(\eta) = \kappa e^{-\kappa}$ is the probability distribution of the last scattering, κ is the differential optical depth, $x = k(\eta_0 - \eta)$ and $j_l(x)$ is the spherical Bessel function.

We obtain the BB-mode correlation spectrum of CMB for primordial gravitational waves for the aforemetioned slow-roll inflationary models. The results are compared with the data from the joint analysis of BICEP2/Keck Array and Planck at 353 GHz [11]. The limit (BK x BK - α BK x P)/(1 - α) at $\alpha = \alpha_{fid} = 0.04$ is evaluated from the auto-spectra and cross-spectra of the joint BICEP2/Keck 150 GHz maps and Planck 353 GHz maps to clean out the dust contribution which is 0.04 times in the BICEP2 band when compared to that in the Planck 353 GHz band; BK x BK indicates the BICEP2/Keck auto-spectra at 150 GHz and BK x P indicates the cross-spectra of BICEP2/Keck maps at 150 GHz and Planck maps at 353 GHz. This combination for the limit is taken after the subtraction of the dust contribution.

The correlation spectrum for the different inflation models is generated using the CAMB code with the tensor spectral index n_T, the tensor-to-scalar ratio r and tensor power spectrum

IOP Conf. Series: Journal of Physics: Conf. Series **883** (2017) 012007 doi:10.1088/1742-6596/883/1/012007

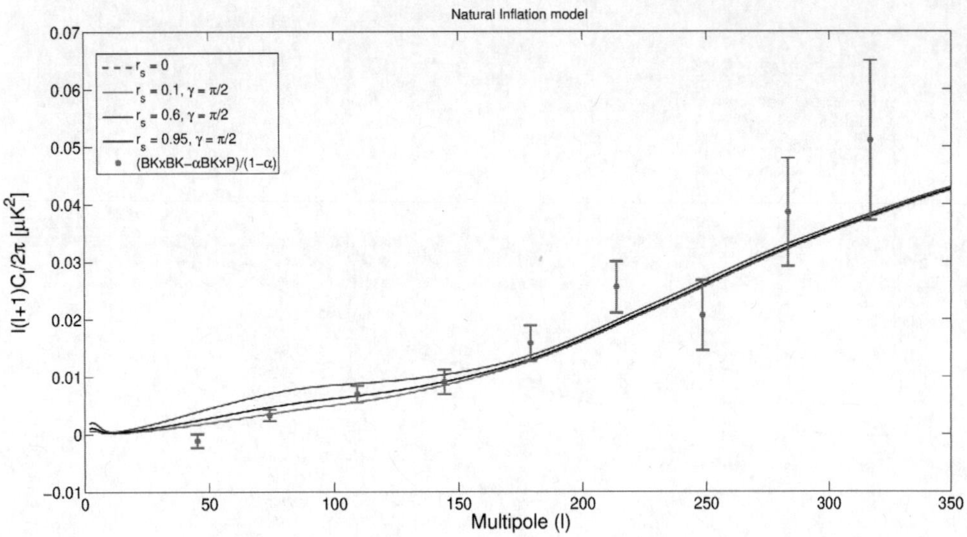

Figure 1. BB-mode angular power spectrum for the Natural inflation model for various values of squeezing parameter.

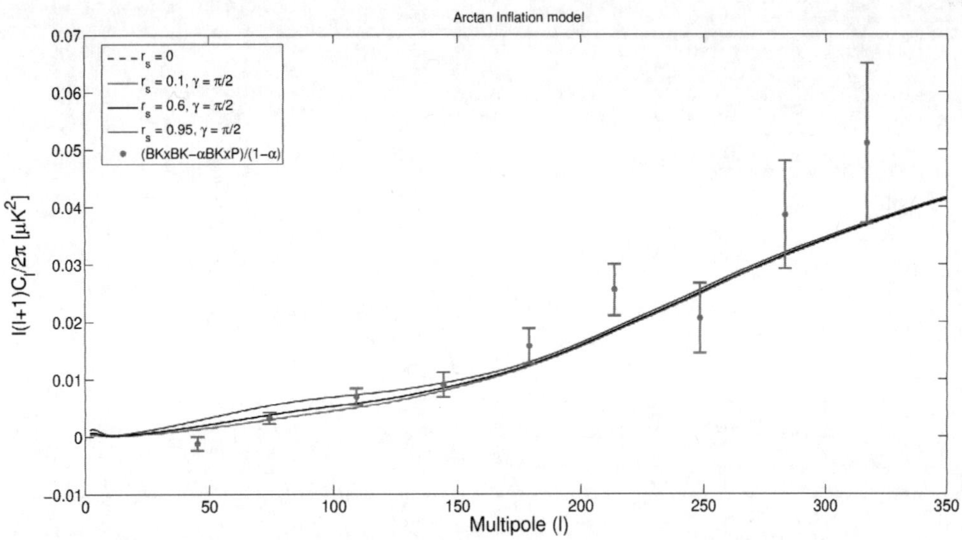

Figure 2. BB-mode angular power spectrum for the Arctan inflation model for various values of squeezing parameter.

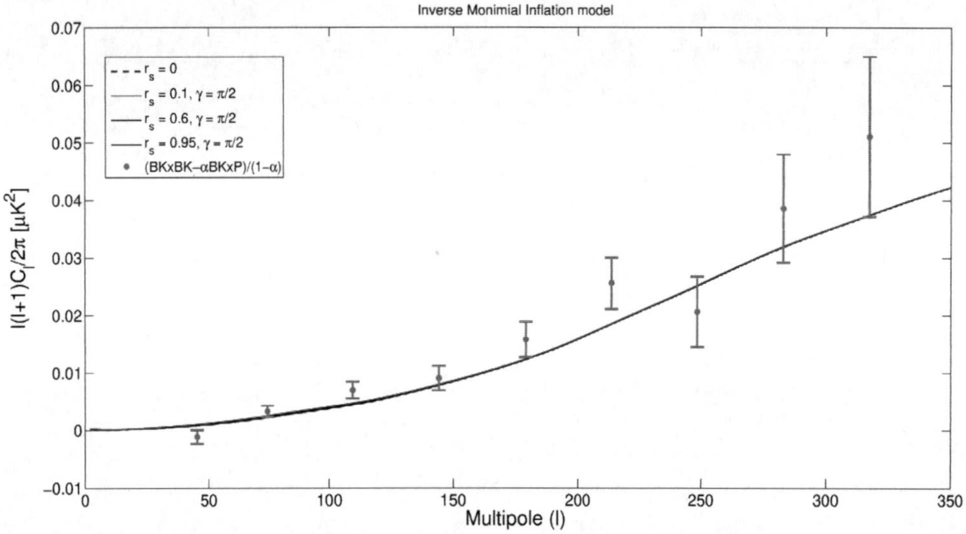

Figure 3. BB-mode angular power spectrum for the Inverse Monomial inflation model for various values of squeezing parameter.

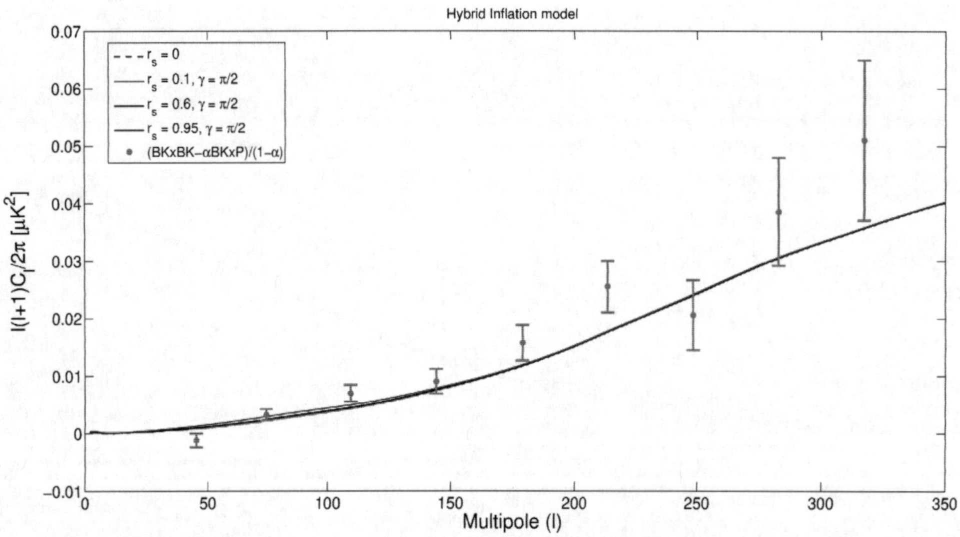

Figure 4. BB-mode angular power spectrum for the Hybrid inflation model for various values of squeezing parameter.

IOP Conf. Series: Journal of Physics: Conf. Series **883** (2017) 012007 doi:10.1088/1742-6596/883/1/012007

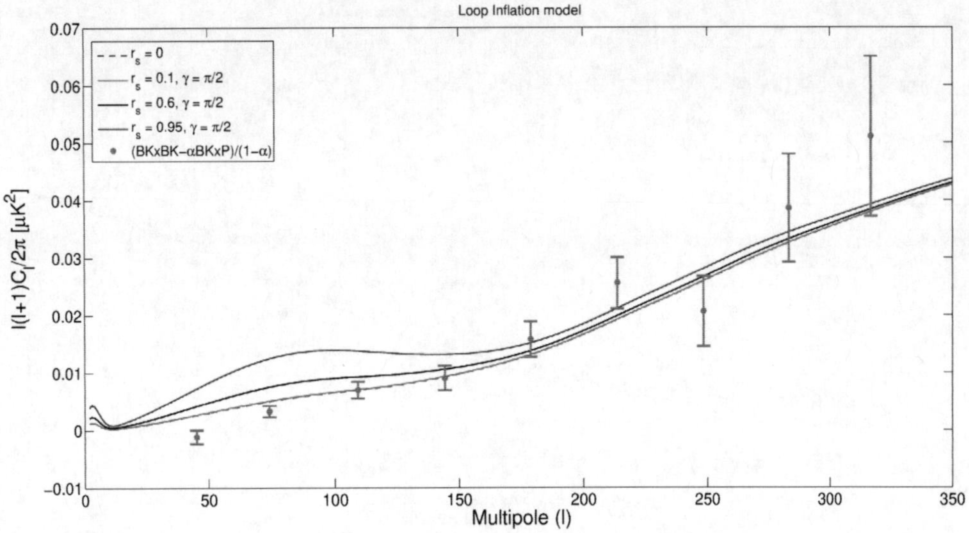

Figure 5. BB-mode angular power spectrum for the Loop inflation model for various values of squeezing parameter.

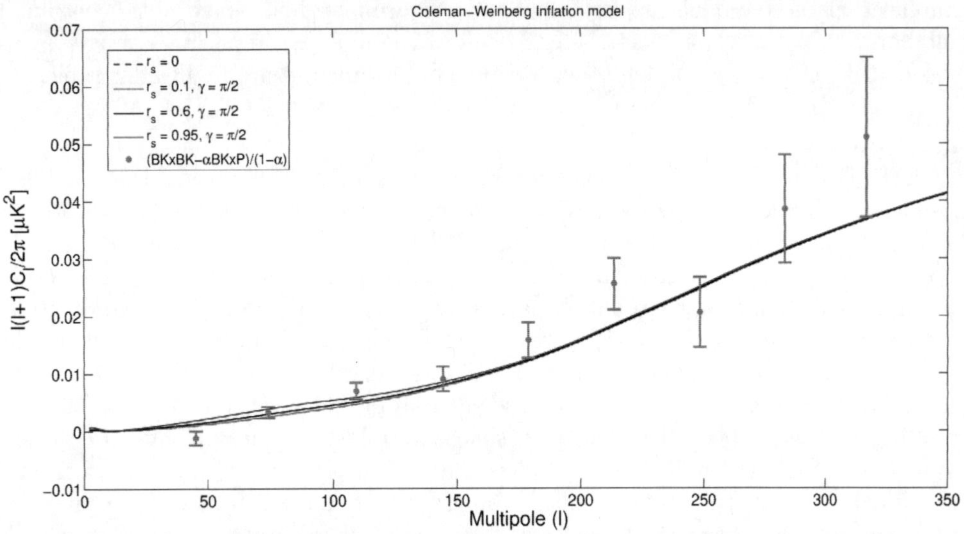

Figure 6. BB-mode angular power spectrum for the Coleman-Weinberg inflation model for various values of squeezing parameter.

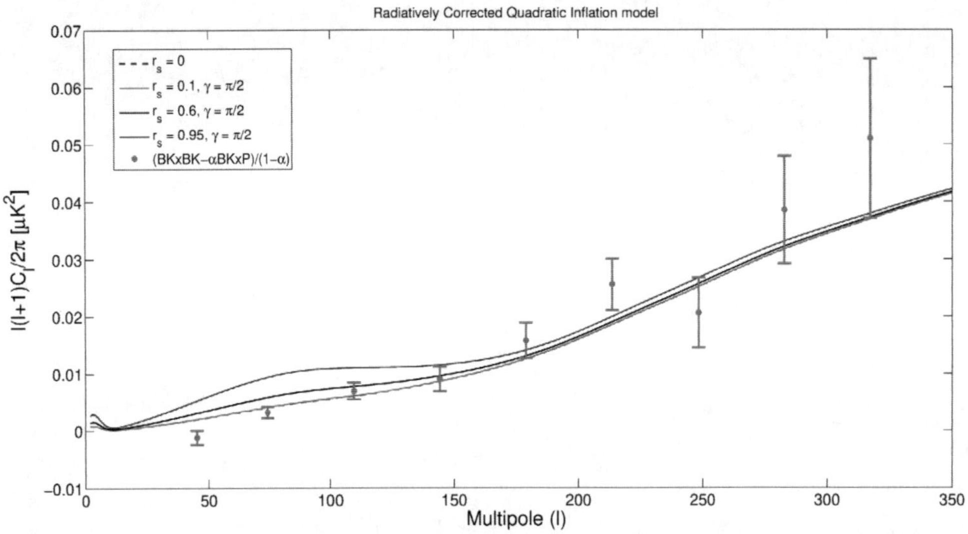

Figure 7. BB-mode angular power spectrum for the Quadratic Chaotic inflation model with radiative corrections for various values of squeezing parameter.

P_T corresponding to each model. For all models, the optical depth is taken to be $\kappa = 0.08$, the pivot wave number for tensor modes is taken as $k_0 = 0.002$ Mpc^{-1} and that for scalar modes is $k_* = 0.05$ Mpc^{-1}.

The BB-mode correlation spectrum for the different inflation models with squeezing effect are obtained and are given in Figs.1, 2, 3, 4, 5, 6 and 7. It can be observed that for inflation models like Natural Inflation model (Fig.1), Arctan Inflation model (Fig.2), Loop Inflation model (Fig.5), Quadratic Chaotic Inflation model with radiative corrections (Fig.7), the angular spectra with all squeezing parameters ($0 \leq r_s \leq 0.95$) do not all lie within the BKP limit, which implies that these models are not ruled out, but are marginally favorable according to the BKP collaboration data, while spectra for the rest of the models (Fig.3, Fig.4, Fig.6), with and without squeezing effect lie completely within the BKP limit and are highly favorable.

6. Conclusion

We studied the BB mode correlation spectrum with squeezing effect for several slow-roll inflation models which are found in agreement with the constraint of the recent collaboration data of BICEP2/ Keck Array at 150 GHz and Planck 353 GHz; we use squeezing effect to help in further constraining these models. Note that the angular power spectrum at higher multipoles (smaller angles) are ignored due to contamination from lensing (B-mode produced due to lensing of E-mode). It is observed that with increase in squeezing parameter for each model, the corresponding GW spectrum also increases. Further, larger the deviation from scale invariance, stronger is the squeezing effect, i.e, for models with larger values of tensor-to-scalar ratio and smaller values of tensor spectral index (n_T), squeezing effect is more prominent, especially at larger values of r_s. All the inflationary models that are studied in the present work are large single field models except the Hybrid Inflation model which is a multi-field inflation model.

The phenomenon of squeezing allows us to treat the resulting state as a stochastic collection of standing waves [37] and thus leads to oscillatory features in the angular spectrum of CMB. Note that for our study, we include anisotropy due to lensing (which acts to smooth out the acoustic peaks in the spectrum on intermediate to small angular scales) in correspondence with the data given by BICEP2/Keck Array and Planck. The primordial GWs, due to squeezing, form a non-stationary background. Hence, the primordial GWs' Gaussian, but non-stationary property can provide a unique signature that could be essential to distinguish a background created from stationary stochastic background by other types of processes which do not produce oscillatory features in the C_l multipole.

Acknowledgments

N M acknowledges the financial support of Council of Scientific and Industrial Research (CSIR), New Delhi, and also would like to thank Asia Pacific Center for Theoretical Physics (APCTP), Korea for financial support for the IF-YITP GR+HEP+Cosmo International Symposium VI. P K S acknowledges the financial support of DST-SERB, New Delhi. The authors would like to thank CAMB, and Planck and BICEP2 websites for the data.

References

[1] Brandenberger R H 1999 *Preprint* hep-ph/9910410v1
[2] Linde A D 1990 *Particle Physics and Inflationary Cosmology* (CRC Press, USA)
[3] Linde A D 2008 *Lect. Notes Phys.* **738** p 1
[4] Guth A H 1981 *Phys. Rev. D* **23** pp 347-56
[5] Martin J, Ringeval C and Vennin V 2014 *Phys. Dark Univ.* **5-6** pp 75-235
[6] Grishchuk L P 1993 *Clas. Quant. Grav.* **10** pp 2449-78
[7] Grishchuk L P 1975 *Sov. Phys. JETP* **40** pp 409-15
[8] Grishchuk L P 2001 *Lect. Notes Phys.* **562** pp 167-94
[9] Grishchuk L P and Sidorov Y V 1989 *Class. Quant. Grav.* **6** pp L161-65
[10] Bose S and Grishchuk L P 2002 *Phys. Rev. D* **66** p 043529
[11] Ade P A R, Aghanim N, *et al* 2015 *Phys. Rev. Lett.* **114** p 101301
[12] Barenboim G and Park W I 2015 *Preprint* astro-ph.CO/1509.07132
[13] Schumaker B L 1986 *Phys. Rep.* **135** pp 317-408
[14] Bhattacharya K, Mohanty S and Nautiyal A 2006 *Phys. Rev. Lett.* **97** p 251301
[15] Freese K, Frieman J A and Olinto A V 1990 *Phys. Rev. Lett.* **65** pp 3233-36
[16] Adams F C, Bond J R, *et al* 1993 *Phys. Rev. D* **47** pp 426-55
[17] Drees M and Erfani E 2012 *JCAP* **2012** p 035
[18] Drees M and Erfani E 2012 *Preprint* astro-ph.CO/1205.4012
[19] Huey G and Lidsey J E 2001 *Phys. Lett. B* **514** pp 217-25
[20] Ratra B and Peebles P J E 1988 *Phys. Rev. D* **37** pp 3406-27
[21] Peebles P J E and Ratra B 1988 *Astrophys. J* **325** pp L17-20
[22] Linde A D 1994 *Phys. Rev. D* **49** pp 748-54
[23] Garcia-Bellido J and Linde A D 1998 *Phys. Rev. D* **57** pp 6075-88
[24] Binetruy P and Dvali G 1996 *Phys. Lett. B* **388** pp 241-46
[25] Halyo E 1996 *Phys. Lett. B* **387** pp 43-47
[26] Dvali D 1996 *Phys. Lett. B* **387** pp 471-77
[27] Coleman S and Weinberg E 1973 *Phys. Rev. D* **7** pp 1888-910
[28] Guth A H and Pi S Y 1982 *Phys. Rev. Lett.* **49** pp 1110-13
[29] Linde A D 1982 *Phys. Lett. B* **116** pp 431-35
[30] Linde A D 1983 *Phys. Lett. B* **129** pp 177-81
[31] Senoguz V N and Shafi Q 2008 *Phys. Lett. B* **668** pp 6-10
[32] Hu W 2008 *Preprint* astro-ph.CO/0802.3688v1
[33] Kosowsky A 1999 *New Astron. Rev.* **43** pp 157-68
[34] Lin Y T and Wandelt B D 2006 *Astropart. Phys.* **25** pp 151-66
[35] Seljak U and Zaldariagga M 1997 *Phys. Rev. Lett.* **78** pp 2054-57
[36] Baskaran D, Grishchuk L P and Polnarev A G 2006 *Phys. Rev. D* **74** p 083008
[37] Grishchuk L P 1992 *Proc. Workshop on Squeezed States and Uncertainty relations* (NASA Goddard Space Flight Center) pp 329-38

IOP Conf. Series: Journal of Physics: Conf. Series **883** (2017) 012008 doi:10.1088/1742-6596/883/1/012008

Non-geometric Five-branes in Heterotic Supergravity

Shin Sasaki* **and Masaya Yata**†

**Department of Physics, Kitasato University, Sagamihara 252-0373, Japan*
†Department of Physics, Tokyo Institute of Technology, Tokyo 152-8551, Japan

E-mail: *shin-s(at)kitasato-u.ac.jp, †yata(at)th.phys.titech.ac.jp

Abstract. We study non-geometric five-branes in heterotic supergravity where the first order α'-corrections are present. By performing the α'-corrected T-duality transformations of the heterotic NS5-brane solutions, we obtain the exotic 5^2_2-brane solutions associated with the symmetric, the neutral and the gauge NS5-branes. We find that the symmetric 5^2_2-brane, and the neutral 5^2_2-brane at $\mathcal{O}(\alpha')$, are T-folds while the gauge 5^2_2-brane is not a T-fold. This proceeding is based on the work [1] where the details that are omitted in this manuscript are found.

1. Introduction

Extended objects such as branes play an important role in superstring theories. The U-duality [2], which makes non-trivial connections among consistent superstring theories, relates various branes in each theory. It was shown that M-theory compactified on T^d has the U-duality symmetry group $E_{d(d)}(\mathbb{R})$ in lower dimensions [3, 4, 5]. BPS branes in superstring theories form lower-dimensional multiplets under the U-duality group. For example, when we consider M-theory compactified on T^8, we have the $E_{8(8)}(\mathbb{R})$ BPS point particle multiplet. The higher dimensional origin of parts of these point particles is the ordinary branes wrapped/unwrapped on cycles in T^8. Here, the ordinary branes are waves, F-strings, D-branes, NS5-branes, Kaluza-Klein (KK) branes. However, there are states whose higher dimensional origin does not trace back to the ordinary branes. These states are called exotic states and their higher dimensional origin is known as exotic branes [6]. Among other things, an exotic brane in type II string theories known as the 5^2_2-brane, has been studied intensively in the past. Type II string theories compactified on T^d has T-duality symmetry group $O(d, d, \mathbb{Z})$. The exotic 5^2_2-brane is obtained by performing the T-duality transformations along the transverse directions to the NS5-brane world-volume. A geometry whose monodromy is characterized by the T-duality symmetry group is known as a T-fold. The geometry associated with the 5^2_2-brane is given by the multi-valued supergravity fields and it has non-geometric nature. However, the 5^2_2-brane geometry has the $O(2, 2)$ monodromy. The multi-valuedness of the geometry is interpreted as the O(2,2) T-duality transformation and it is a consistent solution in string theory. This is an example of the T-fold.

The purpose of the work [1] is to study exotic 5^2_2-branes in heterotic supergravity. In the following, we introduce the explicit solutions of the 5^2_2-branes in heterotic supergravity and show their monodromy structures.

2. Heterotic supergravity

A remarkable property of heterotic supergravity is that the Yang-Mills gauge field contributes to the action as the first order α'-correction. The famous anomaly cancellation mechanism and the

IOP Conf. Series: Journal of Physics: Conf. Series **883** (2017) 012008 doi:10.1088/1742-6596/883/1/012008

supersymmetry completion result in the Riemann curvature square term which involves higher derivative corrections in the same order in α' [7]. The ten-dimensional heterotic supergravity action for the bosonic fields at the first order in α' is given by

$$S = \frac{1}{2\kappa_{10}^2}\int d^{10}x\sqrt{-g}e^{-2\phi}\Big[R(\omega) - \frac{1}{3}\hat{H}^{(3)}_{MNP}\hat{H}^{(3)MNP} + 4\partial_M\phi\partial^M\phi + \alpha'\left(\mathrm{Tr}(F_{MN})^2 + (R_{MNAB}(\omega_+))^2\right)\Big]. \tag{1}$$

Here we employ the convention such that $\frac{\kappa_{10}^2}{2g_{10}^2} = \alpha'$ where κ_{10} and g_{10} are the gravitational and the gauge coupling constants in ten dimensions. The Yang-Mills gauge field strength is denoted by F_{MN} and ϕ is the dilaton. The ten-dimensional metric g_{MN} in the string frame is defined through the vielbein as $g_{MN} = \eta_{AB}e_M{}^A e_N{}^B$ where the metric in the local Lorentz frame is $\eta_{AB} = \mathrm{diag}(-1, 1, \ldots, 1)$. The Ricci scalar $R(\omega)$ and the Riemann tensor $R_{MNAB}(\omega_+)$ are constructed from the spin connection $\omega_M{}^{AB}$. Here $M, N, \ldots, A, B, \ldots$ are the curved space-time and the local Lorentz indices. The modified spin connection is defined by

$$\omega_{\pm M}{}^{AB} = \omega_M{}^{AB} \pm \hat{H}^{(3)AB}_M, \tag{2}$$

where $\hat{H}^{(3)AB}_M = e^{NA}e^{PB}\hat{H}^{(3)}_{MNP}$. The modified H-flux $\hat{H}^{(3)}_{MNP}$ is defined by

$$\hat{H}^{(3)}_{MNP} = H^{(3)}_{MNP} + \alpha'\left(\Omega^{\mathrm{YM}}_{MNP} - \Omega^{\mathrm{L}+}_{MNP}\right). \tag{3}$$

Here $H^{(3)}_{MNP}$ is the ordinary field strength of the NS-NS B-field, $\Omega^{\mathrm{YM}}_{MNP}$, $\Omega^{\mathrm{L}+}_{MNP}$ are the Yang-Mills and the Lorentz Chern-Simons terms, respectively.

3. Five-brane solutions and Buscher rule in heterotic theory

The heterotic NS5-brane solutions satisfy the equation of motion derived from the action (1) at $\mathcal{O}(\alpha')$. There are three distinct NS5-brane solutions known as the symmetric, the neutral and the gauge types in heterotic supergravity [8, 9]. They are 1/2 BPS configurations and preserve a half of the sixteen supercharges. They satisfy the following ansatz:

$$\begin{aligned} &ds^2 = \eta_{ij}dx^i dx^j + H(r)\delta_{mn}dx^m dx^n, \\ &\hat{H}^{(3)}_{mnp} = -\frac{1}{2}\varepsilon_{mnpq}\partial_q H(r), \qquad \phi = \frac{1}{2}\log H(r), \\ &r^2 = (x^1)^2 + (x^2)^2 + (x^3)^2 + (x^4)^2, \end{aligned} \tag{4}$$

where the indices $i, j = 0, 5, 6, 7, 8, 9$, $m, n = 1, 2, 3, 4$ represent the world-volume and the transverse directions to the NS5-branes and ε_{mnpq} is the Levi-Civita symbol. The 1/2 BPS condition leads to the self-duality condition for the Yang-Mills gauge field:

$$F_{mn} = \frac{1}{2}\varepsilon_{mnpq}F_{pq}, \tag{5}$$

and other components A_i vanish. These heterotic NS5-branes are characterized by two charges. One is the topological charge k associated with the gauge instantons:

$$k = -\frac{1}{32\pi^2}\int \mathrm{Tr}[F \wedge F], \tag{6}$$

where the integral is defined in the transverse four-space. The other is the charge Q associated with the modified H-flux:

$$Q = -\frac{1}{2\pi^2\alpha'}\int_{S^3}\hat{H}^{(3)}. \tag{7}$$

Here S^3 is the asymptotic three-sphere surrounding the NS5-brane.

IOP Conf. Series: Journal of Physics: Conf. Series **883** (2017) 012008 doi:10.1088/1742-6596/883/1/012008

3.1. Symmetric solution

The most tractable NS5-brane solution in heterotic theory may be the so-called symmetric solution. The Yang-Mills gauge field which satisfies the self-duality condition (5) is given by an instanton solution in four-dimensions. The gauge field takes value in the $SU(2)$ subgroup of the gauge group $SO(32)$ or $E_8 \times E_8$ and the explicit solution is given by [9]:

$$A_m = -\frac{\sigma_{mn}x^n}{r^2 + n\alpha' e^{-2\phi_0}}, \quad n \in \mathbb{Z},$$
$$H(r) = e^{2\phi_0} + \frac{n\alpha'}{r^2}, \tag{8}$$

where σ_{mn} is the self-dual part of the $SO(4)$ Lorentz generator. The solution (8) is nothing but the BPST one-instanton in the non-singular gauge. The constant ϕ_0 is the asymptotic value of the dilaton. Compared with the general BPST instanton solution, the symmetric solution has a fixed finite instanton size $\rho = e^{-\phi_0}\sqrt{n\alpha'}$. This solution has charges $(k, Q) = (1, n)$. A remarkable fact about the symmetric solution is that the dilaton configuration, hence the harmonic function $H(r)$, is obtained through the standard embedding ansatz:

$$(A_m)^{ab} = \omega_{+m}{}^{ab}. \tag{9}$$

Here the $SU(2)$ indices a, b of the gauge field are identified with the indices of the $SU(2)$ subgroup of the local Lorentz group $SO(4)$ in the transverse directions. If the relation (9) holds, the Chern-Simons terms in the modified H-flux (3) cancel out. Therefore the H-flux only comes from the B-field and the Bianchi identity becomes $d\hat{H}^{(3)} = 0$. From the Bianchi identity and the relation between $H(r)$ and $\hat{H}^{(3)}_{mnp}$ in (4), the B-field is determined through the following condition:

$$\partial_m H = \frac{1}{2}\varepsilon_{mnpq}\partial_n B_{pq}. \tag{10}$$

3.2. Neutral solution

The neutral solution is the one with charges $(k, Q) = (0, n)$ [9]. The Yang-Mills gauge field becomes $A_M = 0$ and the harmonic function is given by

$$H(r) = e^{2\phi_0} + \frac{n\alpha'}{r^2}. \tag{11}$$

For this solution, the curvature term in the modified H-flux becomes a higher order in α' and it is neglected. Since the gauge field does not appear in the neutral solution, this is a solution to type II supergravities.

3.3. Gauge solution

The last is the so called gauge solution. The Yang-Mills gauge field is again given by the BPST instanton. As in the case of the neutral solution, the curvature term in the modified H-flux is neglected as it is a higher order in α'. However, the Yang-Mills gauge field still contributes to the H-flux in the gauge solution and the Bianchi identity becomes $d\hat{H}^{(3)} = \alpha' \mathrm{Tr} F \wedge F$. From the Bianchi identity, the harmonic function is determined to be

$$A_m = -\frac{\sigma_{mn}x^n}{r^2 + \rho^2},$$
$$H(r) = e^{2\phi_0} + 8\alpha' \frac{r^2 + 2\rho^2}{(r^2 + \rho^2)^2}. \tag{12}$$

Compared with the symmetric solution (8), the size modulus ρ is not fixed for the gauge solution and it has charges $(k, Q) = (1, 8)$.

4. Exotic five-branes and monodromies

We now compactify the transverse spaces to the five-brane world-volume to T^2 and perform the T-duality transformation of the solutions. Since the gauge field enters into the action (1) as the first order α'-correction, the T-duality transformation in the heterotic supergravity should be modified from the standard Buscher rule. The first order α'-corrections to the Buscher rule in heterotic supergravity have been written down in [10].

It is convenient to introduce the T-duality covariant quantity $\mathcal{H}$ known as the generalized metric. The generalized metric in heterotic theories is determined by utilizing the heterotic supergravity action (1) compactified on T^d [11]. This is given by

$$\mathcal{H} = \begin{bmatrix} G^{-1} & -G^{-1}B \\ BG^{-1} & G - BG^{-1}B \end{bmatrix}, \tag{13}$$

where G is defined by

$$G_{\mu\nu} = g_{\mu\nu} + 2\alpha'\left[\mathrm{Tr}(\omega_{+\mu}\omega_{+\nu}) - \mathrm{Tr}(A_\mu A_\nu)\right], \tag{14}$$

Here μ, ν are the isometry directions. The generalized metric (13) takes the same form in type II supergravities but the second term in (14) is characteristic to heterotic theories. The spin connection term in (14) has been introduced as it enters into the action (1) in the same way as the Yang-Mills gauge field [11].

4.1. Symmetric solution

By performing the T-duality transformation along the transverse spaces, we obtain the following symmetric 5^2_2-brane solution:

$$\begin{aligned} ds^2 &= \eta_{ij}dx^i dx^j + H\left[(dx^1)^2 + (dx^2)^2\right] + \frac{H}{K}\left[(dx^3)^2 + (dx^4)^2\right], \\ \phi &= \frac{1}{2}\log\frac{H}{K}, \quad B_{34} = -\frac{\omega}{K}, \quad K = H^2 + \omega^2. \end{aligned} \tag{15}$$

The harmonic function $H(r)$ in $\mathbb{R}^2 \times T^2$ is given by [6]:

$$H(r) = h_0 - \frac{\sigma}{2}\log\frac{r}{\mu}, \tag{16}$$

where $\sigma = \frac{R^2}{2\pi\alpha'}$ and μ is a constant which specifies the region where the solution is valid. The constant h_0 diverges in the large cutoff scale limit $\Lambda \to \infty$. Since the harmonic function only depends on x^1, x^2, the NS-NS B-field does so. The only non-zero component of the B-field is determined as $B_{34} \equiv \omega = -\sigma \tan^{-1}\left(\frac{x^2}{x^1}\right)$. The gauge field $A_M{}^I$ is calculated as

$$\begin{aligned} A_1 &= \frac{1}{2H}(\partial_2 H)T^3, \qquad A_2 = -\frac{1}{2H}(\partial_1 H)T^3, \\ A_3 &= \frac{1}{2HK}\Big((H\partial_2 H + \omega\partial_1 H)T^1 - (H\partial_1 H - \omega\partial_2 H)T^2\Big), \\ A_4 &= -\frac{1}{2HK}\Big((H\partial_1 H - \omega\partial_2 H)T^1 + (H\partial_2 H + \omega\partial_1 H)T^2\Big), \\ A_i &= 0. \end{aligned} \tag{17}$$

Here T^I are the $SU(2)$ generators. We find that (17) satisfies the vortex-like equation in two dimensions. We also find that the solution satisfies the standard embedding condition up to a gauge transformation:

$$\mathrm{Tr}F_{MN}F^{MN} = R_{MNAB}(\omega_+)R^{MNBA}(\omega_+). \tag{18}$$

The monodromy structure of the heterotic 5^2_2-brane is characterized by the generalized metric $\mathcal{H}(\theta)$. When we go around the center of the 5^2_2-brane and come back to the original point, namely if the angular position changes as $\theta = \arctan(x^2/x^1) = 0 \to 2\pi$, then the generalized metric for the symmetric 5^2_2-brane solution is evaluated to be

$$\mathcal{H}(2\pi) = O^t \mathcal{H}(0) O, \tag{19}$$

where

$$O = \begin{bmatrix} -i\tau_2 & 0 \\ 2\pi\sigma \mathbf{1}_2 & -i\tau_2 \end{bmatrix} \in O(2,2). \tag{20}$$

This implies that the metric is not single-valued (and non-geometric) but the monodromy for the solution is given by the $O(2,2)$ T-duality transformation. Therefore, we conclude that although the symmetric 5^2_2-brane solution is non-geometric, it is a T-fold and a consistent solution to heterotic string theories.

4.2. Neutral solution

As we have claimed before, the modified spin connection ω_+ is in the $\mathcal{O}(\alpha')$ for the neutral solution. Therefore it is negligible in G_{MN} in the heterotic Buscher rule. After some calculations, the metric, the dilaton and the NS-NS B-field are given by those for the type II 5^2_2-brane solution. This is a conceivable result since the neutral solution does not involve gauge field anymore and it is a solution in type II theory. We stress that although the symmetric solution is an exact solution, the neutral solution is a perturbative solution in the α' expansion. The generalized metric for the neutral solution is given that for the symmetric solution and the monodromy structure is the same with the symmetric case. Therefore we find that the neutral 5^2_2-brane is a T-fold at least at $\mathcal{O}(\alpha')$ in heterotic theories.

4.3. Gauge solution

After performing the T-duality transformation, we find the following gauge 5^2_2-brane solution:

$$\begin{aligned}
ds^2 &= \eta_{ij} dx^i dx^j + \mathcal{I}[(dx^1)^2 + (dx^2)^2] + \frac{\mathcal{I}}{e^{4\phi_0} + \Theta^2}[(dx^3)^2 + (dx^4)^2], \\
B_{34} &= -\frac{\Theta}{e^{4\phi_0} + \Theta^2}, \quad \phi = \frac{1}{2}\log(\frac{\mathcal{I}}{e^{4\phi_0} + \Theta^2}), \quad \mathcal{I} = e^{2\phi_0} - \frac{\alpha'\tilde{\sigma}^2}{2r^2\big(\tilde{h}_0 - \frac{\tilde{\sigma}}{2}\log(r/\mu)\big)^2}, \\
A_1 &= \frac{-\tilde{\sigma}x^2}{4r^2\big(\tilde{h}_0 - \frac{\tilde{\sigma}}{2}\log(r/\mu)\big)} T^3, \quad A_2 = \frac{\tilde{\sigma}x^1}{4r^2\big(\tilde{h}_0 - \frac{\tilde{\sigma}}{2}\log(r/\mu)\big)} T^3, \\
A_3 &= \frac{\tilde{\sigma}}{4r^2\big(\tilde{h}_0 - \frac{\tilde{\sigma}}{2}\log(r/\mu)\big)(e^{4\phi_0} + \Theta^2)} \Big(e^{2\phi_0}(x^1T^2 - x^2T^1) + \Theta(x^1T^1 + x^2T^2)\Big), \\
A_4 &= \frac{\tilde{\sigma}}{4r^2\big(\tilde{h}_0 - \frac{\tilde{\sigma}}{2}\log(r/\mu)\big)(e^{4\phi_0} + \Theta^2)} \Big(e^{2\phi_0}(x^1T^1 + x^2T^2) + \Theta(x^1T^2 - x^2T^1)\Big), \\
A_i &= 0,
\end{aligned} \tag{21}$$

where Θ, $\tilde{\sigma}, \tilde{h}_0$ are constants. The gauge field satisfies the vortex-like equation which is a reminiscent of the self-duality equation (5). In the symmetric and the neutral 5^2_2-brane solutions, there was the function ω which contains explicit angular coordinate on the x^1x^2-plane. In the gauge 5^2_2-brane solution, the B-field is governed by a parameter Θ instead of ω. However, since Θ is a constant parameter, the gauge 5^2_2-brane solution is completely determined by single-valued functions and it is a geometric solution. The function $\mathcal{I}$ becomes negative at a finite value of r.

This is the same situation in the case of the gauge KK5-brane and, unfortunately, its physical meaning is still obscure.

For the gauge 5^2_2-brane, the gauge field contributes to the generalized metric through (14) at $\mathcal{O}(\alpha')$. However, since all the fields in the gauge solution do not depend on the angle θ in the two-dimensional base space, they do not inherit multi-valuedness of the geometry. Therefore the monodromy becomes trivial. This can be seen by evaluating the generalized metric for example in $\Theta = 0$ gauge. In this gauge, we have

$$\mathcal{H} = \begin{bmatrix} G^{-1} & 0 \\ 0 & G \end{bmatrix}, \qquad G = e^{-2\phi_0}\mathbf{1}_2. \tag{22}$$

This implies $\mathcal{H}(2\pi) = \mathcal{H}(0)$. Therefore we concludes that the gauge 5^2_2-brane does not exhibit non-geometric nature.

5. Conclusion

In this proceeding, we show the explicit solutions of the exotic 5^2_2-branes in heterotic supergravity. A specific feature of heterotic supergravity is the Yang-Mills gauge sector which appears in the linear order in the α'-corrections. There are also higher derivative corrections of the curvature square term in the same order in α'. The three different half BPS five-brane solutions in this α' order are known. They are the symmetric, the neutral and the gauge NS5-brane solutions. These are distinguished by the topological charge of instantons of Yang-Mills gauge field and the charge associated with the modified H-flux.

We then introduce $U(1)$ isometries to the heterotic five-brane solutions and perform the T-duality transformations. The resulting solutions are the exotic 5^2_2-branes in heterotic theory. For the symmetric solution, the metric, B-field and the dilation are given by that of the 5^2_2-brane in type II theory. The gauge field satisfies the vortex-like equation in two dimensions. We found that the standard embedding condition is satisfied up to a gauge transformation. For the neutral solution, we found that the neutral 5^2_2-brane is the same with the one in type II theory. They exhibit a non-geometric nature due to the multi-valuedness of the B-field. For the gauge 5^2_2-brane solution, we found that the fields do not show the multi-valuedness and they remain geometric. The monodromies of the three different 5^2_2-branes are also studied. We calculated the generalized metric for these solutions. We found that the symmetric and the neutral 5^2_2-branes have a monodromy given by the $O(2,2)$ T-duality transformation. Therefore they are T-folds. On the other hand, the gauge solution does not show the nature of a T-fold. Details that are not shown in this manuscript are found in the paper [1].

Acknowledgments

The work of S. S. is supported in part by Kitasato University Research Grant for Young Researchers.

References

[1] Sasaki S and Yata M 2016 *JHEP* **1611** 064
[2] Hull M C and Townsend P K 1995 *Nucl.Phys.B* **438** 109
[3] Elitzur S, Giveon A, Kutasov D and Rabinovici E 1998 *Nucl. Phys. B* **509** 122
[4] Obers N A and Pioline B 1999 *Phys.Rept.* **318** 113
[5] Blau M and O'Loughlin M 1998 *Nucl. Phys. B* **525** 182
[6] de Boer J and Shigemori M 2010 *Phys. Rev. Lett.* **104** 251603, 2013 *Phys.Rept.* **532** 65
[7] Bergshoeff E and de Roo M 1989 *Phys.Lett.B* **218** 210, *Nucl.Phys.B* **328** 439
[8] Strominger A 1990 *Nucl.Phys.B* **343** 167
[9] Callan C G Jr., Harvey J A and Strominger A 1991 *Nucl.Phys.B* **359** 611, *Nucl.Phys.B* **367** 60
[10] Bergshoeff E, Janssen B and Ortin T 1996 *Class.Quant.Grav.* **13** 321
[11] Hohm O, Sen A and Zwiebach B 2015 *JHEP* **1502** 079

IOP Conf. Series: Journal of Physics: Conf. Series **883** (2017) 012009 doi:10.1088/1742-6596/883/1/012009

The theory of Minimal Massive Gravity

Antonio De Felice

Center for Gravitational Physics, Yukawa Institute for Theoretical Physics, Kyoto University, 606-8502, Kyoto, Japan

E-mail: antonio.defelice@yukawa.kyoto-u.ac.jp

Abstract. Here we give a short summary of the the theory of minimal massive gravity. This theory, by construction, possesses only two tensor modes, exactly as in General Relativity. This theory is able to reconcile the original background of dRGT – which was proven to possess a ghost – with a stable homogeneous and isotropic cosmological evolution. The theory is constructed by imposing two constraints in a non-linear, background-independent way. The phenomenology of the theory is also briefly discussed.

1. Introduction

Recently the construction of a Lagrangian which was able to give a non-zero mass to the graviton, has attracted many people to try to connect its value to the tiny energy scale which is supposed to the the source of the measured acceleration of our universe.

Unfortunately, this same model proved to lead, in general, to unstable homogeneous and isotropic backgrounds, so that in order to do cosmology with massive gravity one had to renouce either to isotropy or to homogeneity.

In order to try to reconcile a massive graviton to the standard Friedmann-Lemaître-Robertson-Walker (FLRW) background, several authors tried to introduce more degrees of freedom, e.g. by coupling the gravity sector to a scalar field.

Here we discuss a recently proposed theory, which acts in the opposite way, by reducing degrees of freedom. In fact, the theory is built in order to remove all the degrees of freedom of massive theory but the tensor modes. In this way, both scalar and vector modes, in vacuum, are not dynamical leading to the same number of degrees of freedom as in General Relativity (GR). However, the tensor modes acquire a non-zero mass, leading, in general, to a different phenomenology than GR.

2. The model

In order to build up the Lagrangian of the theory we first need to give a three-dimensional fiducial metric as

$$\tilde{\gamma}_{ij} = \delta_{IJ}\, E^I{}_i\, E^J{}_j\,, \tag{1}$$

where $E^I{}_i$ is the fiducial 3-vielbein, together with a second three-tensor as in

$$\tilde{\zeta}^i{}_j = \frac{1}{M}\, E_L{}^i\, \dot{E}^L{}_j\,, \tag{2}$$

Both these fields are considered to be given external fields (as long as we make use of the unitary gauge for the Stückelberg fields). In addition to these fields we need the three-dimensional physical metric

$$\gamma_{ij} = \delta_{IJ}\, e^I{}_i\, e^J{}_j\,, \tag{3}$$

where $e^I{}_i$ is the physical 3-vielbein. After introducing $\tilde{\gamma}_{ij}$, $\tilde{\zeta}^i{}_j$, and γ_{ij}, we can define

$$\kappa^m{}_l\,\kappa^l{}_n = \tilde{\gamma}^{ms}\gamma_{sn}\,, \qquad k^m{}_j\,\kappa^j{}_n = \delta^m{}_n\,, \tag{4}$$

so that $k^m{}_j$ is the inverse tensor of $\kappa^j{}_n$.

Out of these quantities, we introduce the following 3-tensor

$$\Theta^{ij} = \frac{\sqrt{\tilde{\gamma}}}{\sqrt{\gamma}}\{c_1\,(\gamma^{il}\kappa^j{}_l + \gamma^{jl}\,\kappa^i{}_l) + c_2\,[\kappa\,(\gamma^{il}\kappa^j{}_l + \gamma^{jl}\kappa^i{}_l) - 2\tilde{\gamma}^{ij}]\} + 2c_3\gamma^{ij}\,, \tag{5}$$

which is proportional to the difference between the values of the canonical momenta π^{ij} between this theory and GR.

Having defined all the building blocks of the theory we can write down the Lagrangian in the metric formalism as

$$\begin{aligned}
S &= \frac{M_{\rm P}^2}{2}\int d^4x N\sqrt{\gamma}[{}^{(3)}R + K^{ij}K_{ij} - K^2] + \frac{M_{\rm P}^2}{2}\sum_{i=1}^{4}\int d^4x\mathcal{L}_i \\
&+ \frac{M_{\rm P}^2}{2}\int d^4x N\sqrt{\gamma}\left(\frac{m^2N\lambda}{4N}\right)^2\left(\Theta_{ij}\,\Theta^{ij} - \frac{1}{2}\Theta^2\right) \\
&- \frac{M_{\rm P}^2}{2}\int d^4x\sqrt{\gamma}[\lambda\mathcal{C}_0 - (D_n\lambda^i)\mathcal{C}^n{}_i] + S_{\rm matter}\,, &(6)\\
\mathcal{L}_1 &= -m^2\,c_1\,\tilde{a}^3\,(N + M\kappa)\,, &(7)\\
\mathcal{L}_2 &= -\frac{1}{2}\,m^2\,c_2\,\tilde{a}^3\,(2N\kappa + M\kappa^2 - M\kappa^i{}_j\kappa^j{}_i)\,, &(8)\\
\mathcal{L}_3 &= -m^2\,c_3\,\sqrt{\gamma}\,(M + Nk)\,, &(9)\\
\mathcal{L}_4 &= -m^2\,c_4\,\sqrt{\gamma}\,N\,, &(10)\\
\mathcal{C}_0 &= \frac{1}{2}\,m^2\,MK_{ij}\Theta^{ij} - m^2M\left\{\frac{\sqrt{\tilde{\gamma}}}{\sqrt{\gamma}}\,[c_1\zeta + c_2(\kappa\zeta - \kappa^m{}_n\tilde{\zeta}^n{}_m)] + c_3k^m{}_n\tilde{\zeta}^n{}_m\right\}, &(11)\\
\mathcal{C}^n{}_i &= -m^2M\left\{\frac{\sqrt{\tilde{\gamma}}}{\sqrt{\gamma}}\left[\frac{1}{2}\,(c_1 + c_2\kappa)(\kappa^n{}_i + \gamma^{nm}\kappa^l{}_m\gamma_{li}) - c_2\kappa^n{}_l\kappa^l{}_i\right] + c_3\delta^m{}_i\right\}, &(12)\\
S_{\rm matter} &= -\int d^4x\,[\sqrt{-g}\,\rho(n,s) + J^\mu(\partial_\mu\ell + \theta\partial_\mu s + A_1\partial_\mu B_1 + A_2\partial_\mu B_2)]\,. &(13)
\end{aligned}$$

Here, we have introduced two Lagrange multipliers, one 3-scalar λ, and one 3-vector λ^i in order to implement four constraints. These constraints are the necessary ones neeeded to suppress the scalar and vector perturbations of dRGT without spoling its self-accelerating FLRW background solution. Notice that we have supposed here $M = M(t)$, and $E^I{}_j = \tilde{a}(t)\,\delta^I{}_j$. Furthermore N is the lapse function, K_{ij} the extrinsic curvature (K being its trace), D_n represents the covariant derivative with respect to the physical 3D metric, γ_{ij}.

Finally, we have also introduced the matter fields in terms of a perfect fluid, whose action is written in the Schutz-Sorkin form, for which n the number density is written as $n = (J^\alpha J^\beta g_{\alpha\beta}/g)^{1/2}$, and s is the entropy per particle. The fields $A_{1,2}$ and $B_{1,2}$ are scalar fields out of which one can study the vector modes for such a fluid.

3. Phenomenology

In terms of the Hubble factor $H = \dot{a}/aN$ and $X = \tilde{a}/a$, then the background exactly coincides with the one of dRGT theory, namely

$$3H^2 = \frac{m^2}{2}\,(c_4 + 3c_3\,X + 3c_2X^2 + c_1X^3)\,, \tag{14}$$

and two branches still exist as solutions of the following equation

$$(c_3 + 2c_2X + c_1X^2)\,(\dot{X} + NHX - MH) = 0\,. \tag{15}$$

As for the perturbation, the self-accelerating branch, for which $c_3 + 2c_2X + c_1X^2 = 0$, then the dynamics of both scalar and vector modes exactly coincides with the one of GR, whereas the tensor modes acquire a non-zero mass given by

$$\mu^2 = \frac{1}{2}\,m^2\,X\,[c_2X + c_3 + r\,X\,(c_1X + c_2)]\,. \tag{16}$$

However, for the normal branch, the other solution of Eq. (15), we have that the scalar modes for the perfect fluid (modeling dust), gets modified as the dynamics for the $\delta_m = \delta\rho/\rho$ in comoving matter gauge reads as

$$\delta_m'' + \left(2 - \frac{3}{2}\,\Omega_m\right)\delta_m' + \left(\frac{9\theta Y\Omega_m}{2(\theta Y - 2)^2} + \frac{3}{\theta Y - 2}\right)\Omega_m\,\delta_m = 0\,, \tag{17}$$

where $\theta = \mu^2/H_0^2$, and $Y = H_0^2/H^2$.

A non-trivial phenomenology is then predicted to exist for the normal branch, as $G_{\rm eff}$ gets modified. This modification tends to disappear as soon as $Y \ll 1$, i.e. at early times and/or high densities.

4. Conclusions

The theory described here succeeds to build up a theory with a massive graviton with only two tensor degrees of freedom. This theory has non-trivial self-accelerating cosmological solutions which are stable and viable.

Acknowledgments

ADF was supported by JSPS KAKENHI Grant Numbers 16K05348, 16H01099.

References

[1] A. De Felice and S. Mukohyama, *Minimal theory of massive gravity*, Phys. Lett. B **752**, 302 (2016).
[2] A. De Felice and S. Mukohyama, *Phenomenology in minimal theory of massive gravity*, JCAP **1604**, no. 04, 028 (2016).
[3] A. De Felice and S. Mukohyama, *Graviton mass reduces tension between early and late time cosmological data*, arXiv:1607.03368 [astro-ph.CO].

IOP Conf. Series: Journal of Physics: Conf. Series **883** (2017) 012010 doi:10.1088/1742-6596/883/1/012010

Nambu Geometry in Quantum Hall Effect and Topological Insulator

Kazuki Hasebe

Sendai National College of Technology, Ayashi, Sendai 989-3128, Japan

E-mail: khasebe@sendai-nct.ac.jp

Abstract. This short review is a contribution to the conference proceeding of IF-YITP Symposium VI, 2016. We discuss how Nambu geometry emerges in the context of higher dimensional quantum Hall effect or A-class topological insulators [1].

1. Introduction

Non-commutative geometry (NCG) is a promising framework of quantum geometry to extend concepts of classical geometry. While the classical geometry is formulated on infinitely divisible spaces, it is considered that quantum geometry consists of indivisible finite volume elements. A natural scale of quantum space-time will be the Planck scale,

$$\ell_P = \sqrt{\frac{\hbar G}{c^3}} \approx 1.6 \times 10^{-35}[\text{m}]. \tag{1}$$

As the Einstein's theory of general relativity is based on Riemann geometry, the quantum geometry needs a new mathematical framework of quantum space-time. NCG is expected to play the role of such mathematical formulation. Interestingly, NCG also appears effectively in the context of quantum Hall effect (QHE). In QHE, the electrons undergo cyclotron motion with the cyclotron radius,

$$\ell_B = \sqrt{\frac{\hbar}{eB}} \approx 8.1 \times 10^{-9}[\text{m}] \quad \text{for } B = 10[\text{T}]. \tag{2}$$

Though the cyclotron scale is typically 10^{27} times much larger than the Planck scale, QHE realizes the NCG framework due to the level projection [2, 3, 4]. Concretely, on 2D plane in magnetic field, the electron coordinates of cyclotron motion realizes the Heisenberg algebra of NC plane

$$[X, Y] = i{\ell_B}^2, \tag{3}$$

and similarly on a two-sphere in monopole background, electron coordinates satisfy the the $SU(2)$ algebra of fuzzy two-sphere [5, 6, 7]

$$[X_i, X_j] = i\alpha\epsilon_{ijk}X_k. \tag{4}$$

Here both ℓ_B and $\alpha = 2r/I$ (r is the radius of sphere, and $I/2$ is the magnetic charge of the monopole) denote constants of dimension of length related to the indivisible volume element.

It is expected that knowledge obtained from the analysis of the QHE leads us to a better understanding of quantum geometry. We will discuss higher D generalizations of the spherical case (4) in the following.

2. Two Ways to Higher D Non-commutative Geometry

Since the fuzzy two-sphere is based on the $SU(2)$ Lie algebra, one natural way to generalize (4) is to adopt higher D Lie algebras [8, 9, 10, 11, 12, 13]. For realization of $2k$-D fuzzy sphere, the $SO(2k+1)$ gamma matrices are adopted as the coordinates X_a $(a = 1, 2, \cdots, 2k+1)$:

$$[X_a, X_b] = i\alpha X_{ab}, \tag{5}$$

where X_{ab} denote the $SO(2k+1)$ operators. Alternatively by using the quantum Nambu bracket, *i.e.*, a generalized commutator for n entities [14, 15]

$$[X_1, X_2, \cdots, X_n] = \mathrm{sgn}(\sigma) X_{\sigma(1)} X_{\sigma(2)} \cdots, X_{\sigma(n)}, \tag{6}$$

we can generalize (4) to represent the fuzzy $2k$-sphere as [16, 17, 18]

$$[X_{a_1}, X_{a_2}, \cdots, X_{a_{2k}}] = i^k \alpha^{2k-1} \epsilon_{a_1 a_2 \cdots a_{2k+1}} X_{a_{2k+1}}. \tag{7}$$

Hereafter, we will consider the quantum Nambu bracket with even entities. (For odd entities, simple use of the quantum Nambu bracket contains critical difficulties [15].) Though the above two algebras, (5) and (7), take superficially different forms, in both cases X_a are given by the $SO(2n+1)$ gamma matrices, and hence (5) and (7) represent the same fuzzy $2k$-sphere.

The fuzzy two-sphere appears physically in the context of the Landau problem on two-sphere [19]:

$$H = -\frac{1}{2M} \sum_{i=1,2,3} {D_i}^2|_{r=1} \tag{8}$$

where D_i are the covariant derivatives whose commutator gives the monopole field strength,

$$[D_i, D_j] = i\epsilon_{ijk} \frac{I}{2r^2} x_k. \tag{9}$$

Here $I/2$ (I: integer) represents the charge of the monopole. The energy eigenvalues of the Hamiltonian are readily obtained as

$$E_n(I) = \frac{1}{2M}(n(n+1) + I(n+\frac{1}{2})), \tag{10}$$

and the degeneracy of the lowest Landau level ($n = 0$) is

$$d_{LLL}(I) = I + 1. \tag{11}$$

The total $SU(2)$ angular momentum operators are constructed as

$$L_i = -i\epsilon_{ijk} x_j D_k + \frac{1}{\alpha} x_i, \tag{12}$$

and in the lowest Landau level, L_i are reduced to

$$L_i \to \frac{1}{\alpha} x_i. \tag{13}$$

Therefore in the lowest Landau level, x_i can be identified with the operator

$$X_i = \alpha L_i \tag{14}$$

and satisfy the fuzzy two-sphere algebra (4). This is the mechanism of the emergent NCG in the context of the lowest Landau level. Heuristically, the NC algebra (4) can be obtained from (9) by the replacement

$$D_i \rightarrow iX_i. \tag{15}$$

For relativistic case, the Dirac operator acts as the relativistic Hamiltonian. In the presence of non-trivial bundle topology of magnetic field, the Dirac Hamiltonian necessarily accommodates the zero-modes with the number

$$\mathrm{ind}(-i\not{D}) = d_{LLL}(I-1) = I, \tag{16}$$

which coincides with the 1st Chern number on the two-sphere

$$c_1 = \frac{1}{2\pi}\int_{S^2} F = I. \tag{17}$$

This demonstrates the index theorem, $\mathrm{ind}(-i\not{D}) = c_1$.

3. Higher D Quantum Hall Effect and Non-Abelian Gauge Field

The higher D study of the QHE was initiated by the work of Zhang and Hu [20]. They utilized the quaternions to generalize the QHE in four-dimensional space. After their proposal, many works have been devoted to the developments of higher D QHE [21, 22] (see Refs.[23, 24] as reviews and references therein). In particular, we explored higher D QHE on arbitrary even D spheres [25] and on supersphere [26, 27]. The constructions are summarized in Fig.1.

2D	4D	2k D	2|2 D
Complex #	Quaternions	Clifford algebra	Grassmann #
U(1)	SU(2)	SO(2k)	*U*(1)
Dirac monopole	Yang monopole	Non-abelian monopole	Landi super-monopole
1st Chern #	2nd Chern #	kth Chern #	1st Chern #
Fuzzy two-sphere	Fuzzy four-sphere	Fuzzy 2k-sphere	Fuzzy super-sphere

Figure 1. Higher D and supersymmetric generalizations of QHE.

In the setup, we adopted the non-Abelian monopoles introduced in Refs.[28, 29, 30, 31, 32]. For QHE on S^6, we used the $SO(6)$ monopole, and in general, for QHE on S^{2k} $SO(2k)$. The $SO(2k)$ monopole gauge fields are explicitly given by

$$A_{\mu=1,2,\cdots,2k} = -\frac{1}{r(r+x_{2k+1})}\sigma_{\mu\nu}x_\nu, \quad A_{2k+1} = 0, \quad (\sigma_{\mu\nu} : SO(2k) \text{ Weyl generators}) \tag{18}$$

and the covariant derivatives are constructed as $D_a = \partial_a + iA_a$. Their commutators yield the field strength,

$$[D_a, D_b] = iF_{ab} \tag{19}$$

IOP Conf. Series: Journal of Physics: Conf. Series **883** (2017) 012010 doi:10.1088/1742-6596/883/1/012010

where

$$F_{\mu\nu} = \frac{1}{r(r+x_{2k+1})}\sigma_{\mu\nu}x_\nu - \frac{1}{r^2}x_{[\mu}A_{\nu]}, \quad F_{\mu,2k+1} = \frac{1}{r^2}(r+x_{2k+1})A_\mu. \tag{20}$$

Using the covariant derivative, we introduce the Landau Hamiltonian on S^{2k} as

$$H = -\frac{1}{2Mr^2}\sum_{a=1}^{2k+1} {D_a}^2|_{r=1}. \tag{21}$$

The Landau levels are explicitly derived as

$$E_n = \frac{1}{2M}(n(n+2k-1) + I(n+\frac{k}{2})) \quad (n = 0, 1, 2, \cdots). \tag{22}$$

For $2k = 4$ and 6, the lowest Landau level ($n = 0$) degeneracies are given by

$$d_{LLL}^{(2k=4)}(I) = \frac{1}{6}(I+1)(I+2)(I+3), \quad d_{LLL}^{(2k=6)}(I) = \frac{1}{360}(I+1)(I+2)(I+3)^2(I+4)(I+5). \tag{23}$$

The non-Abelian monopoles have the non-trivial Chern numbers guaranteed by the homotopy theorem:

$$\pi_{2k-1}(SO(2k)) \simeq \mathbb{Z}. \tag{24}$$

For $k = 2$ and 3 of (20), the Chern numbers

$$c_k = \frac{1}{k!(2\pi)^k}\int_{S^{2k}} \mathrm{tr}(F^k) \tag{25}$$

are calculated as

$$c_2(I) = \frac{1}{6}I(I+1)(I+2), \quad c_3(I) = \frac{1}{360}I(I+1)(I+2)^2(I+3)(I+4). \tag{26}$$

From (23) and (26), we can confirm that the validity of the index theorem in higher D:

$$\mathrm{Ind}(-i\not{D})_I = d_{LLL}(I-1) = c_k(I). \tag{27}$$

More general discussion can be found in Ref.[33].

Recalling the replacement (15), we can find that the commutation relation (19) implies the NCG

$$[X_a, X_b] = i\alpha X_{ab}, \tag{28}$$

which exactly reproduces (5).

4. Topological Insulator and Tensor Gauge Field

We have constructed a higher D QHE by introducing the non-Abelian monopole. While the 2D restriction was relaxed, magnetic field is still necessary for the set-up. However, the condition about the existence of magnetic field can also be relaxed. The strong spin-orbit coupling takes the place of external magnetic field, and the time reversal counterpart of the QHE, called the quantum spin Hall system, was proposed theoretically [34, 35, 36] and subsequently observed experimentally [37]. Since the spin-orbit coupling is not specific to 2D, the spin-orbit coupling can give rise to the 3D analogue of the quantum spin Hall system referred to as the topological insulator (in a narrow sense) [38, 39, 40]. Thus there are 3D and 2D analogues of the QHE, and there may be no wonder if other cousins of QHE exist. Indeed, a comprehensive list of such cousins, *i.e.* topological insulators in a broad sense, was proposed in Refs.[41, 42, 43] according

IOP Conf. Series: Journal of Physics: Conf. Series **883** (2017) 012010 doi:10.1088/1742-6596/883/1/012010

to the random matrix theory classification based on discrete symmetries. Among them, the original 2D QHE belongs to the so-called A-class, and so the higher D QHE can be identified with higher D entity of the A-class. In Refs.[44, 45], the quantum Nambu geometry was shown to appear as the geometry of the A-class topological insulators by level projection. Meanwhile, as discussed above, the higher D QHE is realized in the non-Abelian monopole background. Therefore, there must be underlying relations between the quantum Nambu geometry and the non-Abelian monopole background. We clarify the relations in the remaining sections.

While the non-Abelian generalization is concerned with the internal gauge space of monopole, there is another way to generalize monopole based on external space extension: The internal gauge group is still $U(1)$, but the external indices are added to the gauge field and the gauge field become a tensor-type [46, 47] (such tensor gauge field is called the Kalb-Ramond field [48]). For 5D and 7D, the tensor monopole gauge fields are given by

$$\begin{aligned} C_{abc} &= -\frac{1}{6r^3}I(I+1)(I+2)\biggl(\frac{1}{r+x_5}+\frac{r}{(r+x_5)^2}\biggr)\epsilon_{abcd5}x_d, \\ C_{abcdef} &= -\frac{1}{40r^5}I(I+1)(I+2)^2(I+3)(I+4)\biggl(\frac{1}{r+x_7}+\frac{r}{(r+x_7)^2}+\frac{2}{3}\frac{r^2}{(r+x_7)^3}\biggr)\epsilon_{abcdef7}x_f \end{aligned} \tag{29}$$

and the corresponding field strengths, $G = dC$, are obtained as

$$G_{abcd} = \frac{1}{2r^5}I(I+1)(I+2)\epsilon_{abcde}x_e, \quad G_{abcdef} = \frac{1}{8r^7}I(I+1)(I+2)^2(I+3)(I+4)\epsilon_{abcdefg}x_g. \tag{30}$$

The integrations of G on S^{2k} yield

$$\int_{S^4} G = \frac{1}{6}I(I+1)(I+2), \quad \int_{S^6} G = \frac{1}{360}I(I+1)(I+2)^2(I+3)(I+4). \tag{31}$$

These results coincide with the Chern numbers (26). Even before the integration, we can make an exact map from the non-Abelian monopole gauge field to the tensor monopole gauge field though the Chern-Simons term

$$C^{(2k-1)}[A] = k\int_0^1 dt\ \mathrm{tr}(A(dA+it^2A^2)^{k-1}) \tag{32}$$

and then

$$G^{2k} = \mathrm{tr}\ F^k. \tag{33}$$

One may convince himself by substituting $SU(2)$ and $SO(6)$ monopole gauge fields (18) to the 4D and 6D Chern-Simons terms

$$C^{(3)}[A] = \mathrm{tr}(AdA + \frac{2}{3}iA^3), \quad C^{(5)}[A] = \mathrm{tr}(A(dA)^2 + \frac{3}{2}iA^3dA - \frac{3}{5}A^5), \tag{34}$$

to derive (29). From (33), we have

$$\frac{1}{r^{2k+1}}x_a = \frac{2}{(2k)!c_k}\epsilon_{aa_1a_2\cdots a_{2k-1}a_{2k}}\,\mathrm{tr}(F_{a_1a_2}\cdots F_{a_{2k-1}a_{2k}}). \tag{35}$$

IOP Conf. Series: Journal of Physics: Conf. Series **883** (2017) 012010 doi:10.1088/1742-6596/883/1/012010

5. Quantum Nambu Geometry and Index Theorem

Armed with the observation of the correspondence between the non-Abelian monopole and tensor monopole, we revisit the non-Abelian Landau problem on S^{2k}:

$$H = -\frac{1}{2Mr^2}\sum_{a=1}^{2k+1} D_a D_a|_{r=1} = \frac{1}{2M}\sum_{a>b}{\Lambda_{ab}}^2, \tag{36}$$

where $\Lambda_{ab} = -x_a D_b + i x_b D_a$ denote the covariant angular momentum. The total angular momentum operators are constructed as

$$L_{ab} = \Lambda_{ab} + r^2 F_{ab}. \tag{37}$$

In the lowest Landau level, the covariant derivatives are quenched and from (37) $F_{ab} \sim L_{ab}$. Consequently, (35) gives the relation

$$X_a = \alpha\frac{I}{(2k)!c_k}\epsilon_{aa_1a_2\cdots a_{2k-1}a_{2k}} L_{a_1a_2}\cdots L_{a_{2k-1}a_{2k}}. \tag{38}$$

The S^{2k} coordinates now become the operator given by (38), and X_a satisfy the higher D algebra of the quantum Nambu bracket (7). Thus, through the correspondence between the non-Abelian and tensor monopoles, we have shown that the quantum Nambu geometry realizes as the geometry of the lowest Landau level [Fig.2].

	Quantum Nambu Geometry	Monopole
2D	2-bracket $[X_i, X_j] = i\epsilon_{ijk}X_k$	U(1) monopole
4D	4-bracket $[X_a, X_b, X_c, X_d] = -\epsilon_{abcde}X_e$	SU(2) monopole ⬍ 3-tensor monopole
6D	6-bracket $[X_{a_1}, X_{a_2}, \cdots, X_{a_6}] = -i\epsilon_{a_1\cdots a_7}X_{a_7}$	SO(6) monopole ⬍ 5-tensor monopole
2k-D	2k-bracket $[X_{a_1}, X_{a_2}, \cdots, X_{a_{2k}}] = i^k\epsilon_{a_1\cdots a_7}X_{a_{2k+1}}$	SO(2k) monopole ⬍ (2k-1)-tensor monopole

Figure 2. Correspondence between the quantum Nambu geometries and the non-Abelian or tensor monopole background.

The higher D fuzzy sphere coordinates are essentially the higher D gamma matrices which can be constructed from the low D ones. Similarly, the higher D quantum Nambu algebra incorporates the low D quantum Nambu algebras, for instance,

$$[X_1, X_2, X_3, X_4] = [X_1, X_2][X_3, X_4] - [X_1, X_3][X_2, X_4] + \cdots \tag{39}$$

and

$$[X_1, X_2, X_3, X_4, X_5, X_6] = [X_1, X_2, X_3, X_4][X_5, X_6] - [X_1, X_2, X_3, X_5][X_4, X_6] + \cdots. \tag{40}$$

These algebraic structures manifest a dimensional hierarchy of the higher D QHE: The low D QH liquids condense to form higher D QH liquid [25]. The index theorem also enforces the observation. The index of the Dirac operator (27) counts the number of the quantum states in the lowest Landau level, and the index theorem tells that such quantum states correspond to the number of the finite volume elements on the quantum geometry given by the Chern number. The index of the non-Abelian Dirac operator on the $2k$D sphere behaves as

$$\text{Ind}(-i\not{D}) = d_{LLL}(I-1) \sim I^{\frac{1}{2}I(I+1)} = I \cdot I^2 \cdot I^3 \cdots I^k, \tag{41}$$

and we see the dimensional hierarchy: The right-hand side of (41) implies that the higher D quantum geometry consists of lower D quantum geometries. Such hierarchy is quite similar to the D-brane structure in string theory [49, 50].

6. Summary

We gave a short review about the Nambu geometry in the context of the monopole magnetic field background along with the development of the higher D QHE and topological insulator. The higher D QHE rediscovered in the recent development of the topological insulator naturally realizes the quantum Nambu geometry. As the Nambu bracket formation is equivalent to the Lie bracket formulation for fuzzy spheres, two superficially different non-Abelian and tensor monopoles are shown to be related by the Chern-Simons term. The index theorem manifests itself in the context of higher D QHE and guarantees the dimensional hierarchy of the quantum geometry. While the present review focused on the even D Nambu geometry, related works about odd D Nambu geometry can be found in Refs.[51, 45, 52].

Acknowledgment

I would like to thank the members of Institute of Fundamental Study, Naresuan University, for inviting me to IF-YITP GR-HEP-Cosmo International Symposium VI. I am also grateful to Shin Sasaki and Seiji Terashima for fruitful discussions during the conference. This work was supported by JSPS KAKENHI Grant Number 16K05334 and 16K05138.

References

[1] Kazuki Hasebe, Nucl. Phys. B 886 (2014) 952-1002.
[2] S.M. Girvin and T. Jach, Phys.Rev. B29 (1984) 5617.
[3] S. M. Girvin, A. H. MacDonald, and P. M. Platzman, Phys. Rev. Lett. 54 (1985) 581; Phys. Rev. B33 (1986) 2481.
[4] Z.F. Ezawa, G. Tsitsishvili, K. Hasebe, Phys.Rev. B67 (2003) 125314.
[5] F.A. Berezin, Commun.Math. Phys. 40 (1975) 153-174.
[6] J. Hoppe, MIT PhD Thesis (1982).
[7] J. Madore, Class. Quant. Grav. 9 (1992) 69.
[8] H. Grosse, C. Klimcik, P. Presnajder, Commun.Math.Phys. 180 (1996) 429-438.
[9] D. Kabat, W. Taylor, Adv.Theor.Math.Phys.2 (1998) 181-206.
[10] S. Ramgoolam, Nucl.Phys. B610 (2001) 461-488.
[11] P. M. Ho and S. Ramgoolam, Nucl.Phys.B 627 (2002) 266.
[12] Yusuke Kimura, Nucl.Phys. B637 (2002) 177-198.
[13] Yusuke Kimura, Nucl.Phys. B664 (2003) 512-530.
[14] Yoichiro Nambu, Phys.Rev.D7 (1973) 2405-2412.
[15] Thomas Curtright, Cosmas Zachos, Phys.Rev.D68 (2003) 085001.
[16] M. M. Sheikh-Jabbari, JHEP 0409 (2004) 017.
[17] M. M. Sheikh-Jabbari, M. Torabian, JHEP 0504 (2005) 001.
[18] Joshua DeBellis, Christian Saemann, Richard J. Szabo, J.Math.Phys.51 (2010) 122303.
[19] F.D.M. Haldane, Phys. Rev. Lett. 51 (1983) 605-608.

IOP Conf. Series: Journal of Physics: Conf. Series **883** (2017) 012010 doi:10.1088/1742-6596/883/1/012010

[20] S. C. Zhang and J. P. Hu, Science 294 (2001) 823.
[21] D. Karabali, V. P. Nair, Nucl.Phys. B641 (2002) 533.
[22] B. A. Bernevig, J. P. Hu, N. Toumbas, S. C. Zhang, Phys.Rev.Lett. 91 (2003) 236803.
[23] Kazuki Hasebe, SIGMA 6 (2010) 071.
[24] Dimitra Karabali, V.P. Nair, J.Phys.A39 (2006) 12735-12764.
[25] K. Hasebe and Y. Kimura, Phys.Lett. B 602 (2004) 255.
[26] Kazuki Hasebe, Phys.Rev.Lett. 94 (2005) 206802.
[27] Kazuki Hasebe, Yusuke Kimura Nucl.Phys. B709 (2005) 94-114.
[28] Chen Ning Yang, J. Math. Phys. 19 (1978) 320.
[29] Z. Horvath and L. Palla, Nucl.Phys. B142 (1978) 327.
[30] D. H. Tchrakian, J.Math. Phys. 21 (1980) 166.
[31] Bernard Grossman, Thomas W. Kephart, James D. Stasheff, Commun.Math.Phys. 96 (1984) 431, Erratum-ibid.100 (1985) 311.
[32] C. Saclioglu, Nucl.Phys. B277 (1986) 487.
[33] Brian P. Dolan, JHEP 0305 (2003) 018.
[34] C.L. Kane, E.J. Mele, Phys. Rev. Lett. 95 (2005) 146802.
[35] B. Andrei Bernevig, Shou-Cheng Zhang, Phys. Rev. Lett. 96 (2006) 106802.
[36] B. Andrei Bernevig, Taylor L. Hughes, Shou-Cheng Zhang, Science, 314 (2006) 1757.
[37] Markus Koenig et al, Science, 318 (2007) 766.
[38] Liang Fu and C. L. Kane, Phys. Rev. B 76 (2007) 045302.
[39] Shuichi Murakami, New Journal of Physics 9 (2007) 356.
[40] D. Hsieh, D. Qian, L. Wray, Y. Xia, Y.S. Hor, R.J. Cava, M.Z. Hasan, Nature 452 (2008) 970-974.
[41] Andreas P. Schnyder, Shinsei Ryu, Akira Furusaki, Andreas W. W. Ludwig, Phys. Rev. B 78 (2008) 195125.
[42] A. Kitaev, arXiv:0901.2686.
[43] Shinsei Ryu, Andreas Schnyder, Akira Furusaki, Andreas Ludwig, New J. Phys. 12 (2010) 065010.
[44] B. Estienne, N. Regnault, B. A. Bernevig, Phys. Rev. B 86 (2012) 241104(R).
[45] Titus Neupert, Luiz Santos, Shinsei Ryu, Claudio Chamon, Christopher Mudry, Phys. Rev. B 86 (2012) 035125.
[46] R.I. Nepomechie, Phys. Rev. D 31 (1985) 1921.
[47] Claudio Teitelboim, Phys.Lett. B 167 (1986) 63-68, Phys.Lett. B 167 (1986) 69-72.
[48] Michael Kalb, P. Ramond, Phys.Rev. D 9 (1974) 2273-2284.
[49] P.K. Townsend, hep-th/9507048.
[50] Tsuguhiko Asakawa, Shigeki Sugimoto, Seiji Terashima, JHEP 0302 (2003) 011.
[51] Kazuki Hasebe, Nucl. Phys. B 886 (2014) 681-690.
[52] V.P. Nair, S. Randjbar-Daemi, Nucl.Phys. B679 (2004) 447-463.

IOP Conf. Series: Journal of Physics: Conf. Series **883** (2017) 012011 doi:10.1088/1742-6596/883/1/012011

Robin Gravity

Chethan Krishnan, Shubham Maheshwari and P. N. Bala Subramanian

Centre for High Energy Physics, Indian Institute of Science, Bangalore, India. 560012

E-mail:
chethan.krishnan@gmail.com,shubham.93@gmail.com,pnbalasubramanian@gmail.com

Abstract. We write down a Robin boundary term for general relativity. The construction relies on the Neumann result of arXiv:1605.01603 in an essential way. This is unlike in mechanics and (polynomial) field theory, where two formulations of the Robin problem exist: one with Dirichlet as the natural limiting case, and another with Neumann.

1. Introduction

The Dirichlet boundary term for general relativity was found by York and Gibbons-Hawking[1] long ago, but the Neumann term was only written down recently [2]. See [3, 4] for follow-ups. In this contribution, we will further develop the result of [2] to construct a well-defined Robin variational problem and construct the general Robin boundary term for general relativity. The Neumann term turns out to be essential for the Robin construction in gravity in a way that it is not, in mechanics and field theory. To clarify this, we outline the various constructions systematically. We also briefly comment on aspects of such a term in asymptotically flat and asymptotically AdS spacetimes.

This contribution is a small application of the results in [2, 3, 4]. But the existence of a Robin term for general relativity does not seem to have been appreciated in the literature, so we hope it will be of some use to someone somewhere sometime.

2. Robin Mechanics

Let us start by looking at boundary terms in the simplest setting: particle mechanics. Consider the usual Newton action

$$S_D^p[q] = \int_T dt\ L(q, \dot{q}) = \int_T dt\ \left(\frac{1}{2}\dot{q}^2 - V(q)\right). \tag{1}$$

The superscript p indicates the action is that of a particle, and the subscript D denotes that it leads to a well defined variational problem with *Dirichlet* boundary condition (in the time direction). To restate the well-known, the variation of the action gives

$$\delta S_D^p = -\int_T dt\ \left(\ddot{q} + V'(q)\right)\delta q + (\dot{q}\delta q)\Big|_{\partial T}. \tag{2}$$

IOP Conf. Series: Journal of Physics: Conf. Series **883** (2017) 012011 doi:10.1088/1742-6596/883/1/012011

If one sets $q = \textit{any fixed quantity}$ at the endpoint[1] T, the variational problem becomes well posed, and since we are setting $\delta q|_T = 0$ we call it a Dirichlet problem.

Note that setting $\dot{q}|_T = 0$ in δS^p_D is another way to define a valid variational problem, while *not* demanding[2] that $\delta q = 0$. We will call this the *Special Neumann* boundary condition.

We would like to find a variational principle where holding $\delta\dot{q} = 0$ is well-defined. This is the natural *General Neumann* boundary condition, and to accomplish this we add a boundary term to the action:

$$S^p_N = S^p_D - (q\,\dot{q})\Big|_T, \tag{3}$$

$$\Rightarrow \delta S^p_N = -\int_T dt\,\left(\ddot{q} + V'(q)\right)\delta q - (q\delta\dot{q})\Big|_T. \tag{4}$$

We could restate it in terms of conjugate quantities at the boundary, which leads to a more useful notation later, as

$$(q\,\dot{q})\Big|_T \equiv \pi_T\, q_T, \text{ where } \pi_T \equiv \frac{\delta S^p_D}{\delta q_T}. \tag{5}$$

Note that General Neumann boundary conditions basically mean fixing $\dot{q} = \textit{any fixed value}$, while Special Neumann boundary condition allows only the possibility $\dot{q} = 0$.

Now, let us consider another boundary term that we could add to S^p_D, namely $S^p_1 = \frac{\xi}{2}q^2|_T$. This generalizes the Special Neumann boundary condition and leads to what we will call the *Special Robin* boundary condition. Upon varying $S^p_D + S^p_1$ we get

$$\delta(S^p_D + S^p_1) = -\int_T dt\,\left(\ddot{q} + V'(q)\right)\delta q + \left(\dot{q} + \xi\,q\right)\delta q\Big|_T. \tag{6}$$

If we set $\delta q = 0$, this is still the Dirichlet variational problem. But we can also set $(\dot{q}+\xi q)|_T = 0$, which is the Special Robin boundary condition: holding a linear combination of the position and velocity fixed at the boundary. When $\xi = 0$, this reduces to the Special Neumann boundary condition.

What is the Robin analogue of the General Neumann boundary condition? Lets consider adding one more piece to our General Neumann action[3]:

$$S^p_R = \int_T dt\left(\frac{1}{2}\dot{q}^2 - V(q)\right) - \left(\dot{q}\,q + \frac{\xi}{2}q^2\right)\Big|_T, \tag{7}$$

which upon varying gives

$$\delta S^p_R = -\int_T dt\,\left(\ddot{q} + V'(q)\right)\delta q - q\,\delta\left(\dot{q} + \xi q\right)\Big|_T. \tag{8}$$

The variational problem is well defined by setting $\dot{q} + \xi q = \textit{any fixed value}$. This is the *General Robin* boundary condition. Again, we could phrase the whole thing as

$$\delta S^p_R = \text{eoms} - q_T\,\delta(\pi_T + \xi q_T) = \text{eoms} - q_T\,\delta\left(\frac{\delta S^p_D}{\delta q_T} + \xi q_T\right). \tag{9}$$

[1] We keep track of only one boundary, as it is suffices to make our point.

[2] Demanding both $\delta q = 0 = \dot{q}$ fixes both the function and its derivative at the boundary, constraining dynamics uniquely. This is not what we want from a theory: it should allow dynamics, not uniquely fix it.

[3] It is possible to set up a General Robin boundary problem for particle mechanics, by starting with the Dirichlet action and never going through the Neumann action. However, this approach does not work for gravity and we find that the Neumann action is crucial for the construction of the Robin action for general relativity. We discuss these matters in Appendix A.

IOP Conf. Series: Journal of Physics: Conf. Series **883** (2017) 012011 doi:10.1088/1742-6596/883/1/012011

The Dirichlet problem can be understood as a variational problem with the position of the particle held arbitrary and fixed at T, and the General Neumann problem to be a variational problem with the momentum at T held arbitrary and fixed. The General Robin boundary condition is analogously to be thought of as holding some linear combination of the position and momentum held arbitrary and fixed at T. The Dirichlet [1] and General Neumann problem [2] for general relativity are solved, here we would like to fill the gap and formulate the General Robin problem for gravity. There is a bit of a subtlety in this compared to the particle mechanics case (see footnote 3).

But before getting to gravity, we consider the field theory case which is essentially just a fancy rewriting of the particle mechanics case.

3. Robin Field Theory

We will start with the action for a scalar field living in a D-dimensional manifold $(\mathcal{M}, g)$, which again is automatically a Dirichlet action, where we hold the field to be at some fixed value at the boundary

$$S_D[\phi] = \int_{\mathcal{M}} d^D x\sqrt{-g}\left(-\frac{1}{2}g^{\mu\nu}\partial_\mu\phi\partial_\nu\phi - V(\phi)\right), \tag{10}$$

$$\Rightarrow \delta S_D[\phi] = \int_{\mathcal{M}} d^D x\left(\partial_\mu(\sqrt{-g}g^{\mu\nu}\partial_\nu\phi) - \frac{\partial V(\phi)}{\partial\phi}\right)\delta\phi - \int_{\partial\mathcal{M}} d^{D-1}x\sqrt{|\gamma|}n_\mu g^{\mu\nu}\partial_\nu\phi\,\delta\phi \tag{11}$$

where γ is the metric on the boundary $\partial\mathcal{M}$ of $\mathcal{M}$, and n_μ is the outward drawn normal to the boundary. The standard procedure, as mentioned earlier is to set the Dirichlet boundary condition $\delta\phi = 0$, which leads to a well defined variational problem. The Special Neumann case is obtained from the same action while setting the directional derivative $\partial_n\phi \equiv n_\mu g^{\mu\nu}\partial_\nu\phi = 0$ instead. As before, if we work with

$$S_D[\phi] + \int_{\partial\mathcal{M}} d^{D-1}x\sqrt{|\gamma|}\,\frac{\xi}{2}\phi^2 \tag{12}$$

it leads to the Special Robin variational problem with $\partial_n\phi + \xi\phi = 0$ at the boundary. To get the General Neumann action, we take by direct analogy

$$S_N[\partial_n\phi] = S_D[\phi] + \int_{\partial\mathcal{M}} d^{D-1}x\sqrt{|\gamma|}(n_\mu g^{\mu\nu}\partial_\nu\phi)\,\phi \tag{13}$$

$$\Rightarrow \delta S_N[\partial_n\phi] = \text{eom} + \int_{\partial\mathcal{M}} d^{D-1}x\sqrt{|\gamma|}\,\delta(\partial_n\phi)\,\phi. \tag{14}$$

The variational problem here is well defined by holding $\delta(\partial_n\phi) = 0$. The scalar field theory can also be well posed as a General Robin boundary problem:

$$S_R[\phi] = S_D[\phi] + \int_{\partial\mathcal{M}} d^{D-1}x\sqrt{|\gamma|}\left(\phi\,\partial_n\phi + \frac{\xi}{2}\phi^2\right) = S_N[\phi] + \int_{\partial\mathcal{M}} d^{D-1}x\sqrt{|\gamma|}\,\frac{\xi}{2}\phi^2, \tag{15}$$

which will lead to holding $\partial_n\phi + \xi\phi = \textit{any fixed value}$ at $\partial\mathcal{M}$. We worked with the scalar for simplicity, but this generalizes trivially to the gauge field as well.

4. Robin Gravity

We can now proceed to look for a boundary term that gives a consistent Robin boundary problem for gravity. Let us, as usual, start with the Einstein-Hilbert action on a D-dimensional manifold $(\mathcal{M}, g)$ along with the Gibbons-Hawking York boundary term, which leads to Dirichlet gravity

$$S_D = S_{EH} + S_{GHY} = \frac{1}{2\kappa}\int_{\mathcal{M}} d^D x\,\sqrt{-g}(R - 2\Lambda) + \frac{1}{\kappa}\int_{\partial\mathcal{M}} d^{D-1}y\sqrt{|\gamma|}\,\epsilon\,\Theta, \tag{16}$$

where $\kappa = 8\pi G_N$, R is the Ricci scalar and Λ is a cosmological constant. Also, $\gamma_{ij} = g_{\mu\nu}e^\mu_i e^\nu_j$ is the induced metric on the boundary $\partial\mathcal{M}$ and $e^\mu_i = \frac{\partial x^\mu}{\partial y^i}$ projects the bulk coordinates x^μ to the boundary coordinates y^i. The extrinsic curvature of the boundary is given by

$$\Theta_{ij} = \frac{1}{2}(\nabla_\mu n_\nu + \nabla_\nu n_\mu)e^\mu_i e^\nu_j, \tag{17}$$

where n_μ is the outward drawn unit normal to the boundary, and $\epsilon = \pm 1$ distinguishes the boundary between time-like and space-like boundaries respectively.

The variation of Dirichlet action yields

$$\begin{aligned} \delta S_D &= \delta S_{EH} + \delta S_{GHY} \\ &= \frac{1}{2\kappa}\int_{\mathcal{M}} d^D x\,\sqrt{-g}\,(G_{\mu\nu} + \Lambda g_{\mu\nu})\,\delta g^{\mu\nu} - \frac{1}{2\kappa}\int_{\partial\mathcal{M}} d^{D-1}y\sqrt{|\gamma|}\epsilon\left(\Theta^{ij} - \Theta\gamma^{ij}\right)\delta\gamma_{ij}, \end{aligned} \tag{18}$$

where $G_{\mu\nu} = R_{\mu\nu} - \frac{1}{2}g_{\mu\nu}R$ is the Einstein tensor. The variational problem is well defined with the boundary metric held fixed, and we can think of $S_D = S_D[\gamma_{ij}]$ as a functional of the boundary metric.

We can define a canonical conjugate of the boundary metric as

$$\pi^{ij} \equiv \frac{\delta S_D}{\delta\gamma_{ij}} = -\frac{1}{2\kappa}\sqrt{|\gamma|}\epsilon\left(\Theta^{ij} - \Theta\gamma^{ij}\right), \tag{19}$$

using which we can rewrite the variation of S_D in a simpler form

$$\delta S_D = \frac{1}{2\kappa}\int_{\mathcal{M}} d^D x\,\sqrt{-g}\,(G_{\mu\nu} + \Lambda g_{\mu\nu})\,\delta g^{\mu\nu} + \int_{\partial\mathcal{M}} d^{D-1}y\,\pi^{ij}\,\delta\gamma_{ij}. \tag{20}$$

We also note that holding $\pi^{ij} = 0$ here leads to the Special Neumann boundary condition for gravity. This is sometimes described as the Neumann problem for gravity in the literature, even though it is a special case of the general situation.

As was discussed in [2, 3, 4], an action which is well defined in terms of General Neumann boundary condition can be defined as

$$\begin{aligned} S_N &= S_{EH} + S_{GHY} - \int_{\partial\mathcal{M}} d^{D-1}y\,\pi^{ij}\,\gamma_{ij} \qquad (21) \\ &= \frac{1}{2\kappa}\int_{\mathcal{M}} d^D x\,\sqrt{-g}(R - 2\Lambda) + \frac{4-D}{2\kappa}\int_{\partial\mathcal{M}} d^{D-1}y\sqrt{|\gamma|}\epsilon\Theta, \qquad (22) \end{aligned}$$

the variation of which is given by

$$\delta S_N = \frac{1}{2\kappa}\int_{\mathcal{M}} d^D x\,\sqrt{-g}\,(G_{\mu\nu} + \Lambda g_{\mu\nu})\,\delta g^{\mu\nu} + \int_{\partial\mathcal{M}} d^{D-1}y\,\gamma_{ij}\,\delta\pi^{ij}. \tag{23}$$

Here, instead of holding the boundary metric fixed, the quantity π^{ij} is held fixed, letting the boundary metric fluctuate. The quantity π^{ij} is termed boundary stress tensor density (also, sometimes as quasi-local stress tensor density[5]). The Neumann boundary condition can be thought of as looking at solutions holding the boundary stress tensor density fixed, i.e. $S_N = S_N[\pi^{ij}]$.

Now we turn to Special Robin. Adding a boundary term $S_b = 2\zeta\int d^{D-1}y\sqrt{|\gamma|}$ to the Dirichlet action, it is straightforward to again check that we will have a variational problem well defined under the Special Robin boundary condition, $\pi^{ij} + \zeta\sqrt{|\gamma|}\gamma^{ij} = 0$.

IOP Conf. Series: Journal of Physics: Conf. Series **883** (2017) 012011 doi:10.1088/1742-6596/883/1/012011

In order to have the action be a well defined variational problem under General Robin boundary condition, we need to add a boundary term which will ensure that $\pi^{ij} + \xi\sqrt{|\gamma|}\gamma^{ij}$ held arbitrary and fixed[4] leads to a consistent variational problem. To get such an action, we go through the Neumann action like we did in the mechanics and field theory cases. Note that unlike in those cases, in gravity we cannot get to Robin from Dirichlet bypassing Neumann[5]. In other words, going through Neumann is not an option but a necessity in the case of gravity.

In any event, the result is

$$\begin{aligned} S_R &= \frac{1}{2\kappa}\int_{\mathcal{M}} d^D x\,\sqrt{-g}(R-2\Lambda) \\ &\quad + \frac{4-D}{2\kappa}\int_{\partial\mathcal{M}} d^{D-1}y\sqrt{|\gamma|}\epsilon\Theta - \xi(D-3)\int_{\partial\mathcal{M}} d^{D-1}y\sqrt{|\gamma|}. \end{aligned} \tag{24}$$

Varying the action, and using the key relation

$$(D-3)\sqrt{|\gamma|}\gamma^{ij}\delta\gamma_{ij} = 2\,\delta(\sqrt{|\gamma|}\gamma^{ij})\,\gamma_{ij}, \tag{25}$$

we can show that

$$\delta S_R = \frac{1}{2\kappa}\int_{\mathcal{M}} d^D x\,\sqrt{-g}\,(G_{\mu\nu}+\Lambda g_{\mu\nu})\,\delta g^{\mu\nu} - \int_{\partial\mathcal{M}} d^{D-1}y\,\delta(\pi^{ij}+\xi\sqrt{|\gamma|}\gamma^{ij})\,\gamma_{ij}. \tag{26}$$

The action (24) is what we call the Robin action for gravity.

5. Comments

Hamiltonian Formulation

We will now write down the Robin Gravity action in the Hamiltonian formulation. Using the fact that the action in (24) is the same as that of Neumann gravity, except for an additional boundary cosmological constant term, we can directly write down the action in terms of canonical variables for Robin gravity [3]:

$$\begin{aligned} S_R &= \int_{\mathcal{M}} d^D x\left(p^{ab}\dot{h}_{ab} - NH - N_a H^a\right) \\ &\quad + \int_{\mathcal{B}} d^{D-1}y\sqrt{\sigma}\left(N\left(\frac{\varepsilon}{2} - \xi(d-3)\right) - N^a j_a + \frac{N}{2}s^{ab}\sigma_{ab}\right). \end{aligned} \tag{27}$$

We will not elaborate on the (completely standard) notations here, they can be found in, eg. [3].

Asymptotically Flat Space-times

If one goes about naively computing the classical action for any asymptotically flat space-time(AFS), its bound to run into divergences. The usual procedure to deal with in AFS is to do a background subtraction, which involves holding the induced metric at the boundary the same for the background and the datum. For (24), this means that the boundary cosmological constant term drops off and we end up with the same result as what one would get from a pure Neumann boundary term, see [3]. This means that various discussions there on thermodynamics, horizons, etc [6] also immediately apply to the background subtracted case here.

[4] The explicit presence of $\sqrt{|\gamma|}$ is not of much worry, as one can see, π^{ij} is also defined implicitly with the same factor.

[5] See discussion in the Appendix for some elaboration on this.

IOP Conf. Series: Journal of Physics: Conf. Series **883** (2017) 012011 doi:10.1088/1742-6596/883/1/012011

AdS

We will now look at asymptotically AdS_{d+1} spaces[6]. We follow the notations of [4].

The renormalized Neumann action is given by

$$S_N^{ren} = S_D^{ren} - \int_{\partial\mathcal{M}} d^d x\, \pi^{ij}_{ren}\gamma_{ij}, \tag{28}$$

where π^{ij}_{ren} is the renormaized boundary stress tensor density.

The boundary stress tensor is related to π^{ij} as[7]

$$T^{ij} = -\frac{2}{\sqrt{-\gamma}}\pi^{ij} = \frac{1}{\kappa}(\Theta^{ij} - \Theta\,\gamma^{ij}), \tag{29}$$

and the renormalized stress tensor T^{ij}_{ren} is obtained from π^{ij}_{ren} in the same way. The variation of renormalized Neumann action gives

$$\begin{aligned} \delta S_N^{ren} &= \text{eq. of motion } - \int_{\partial\mathcal{M}} d^d x\, \gamma_{ij}\delta\pi^{ij}_{ren} \\ &= \text{eq. of motion } + \frac{1}{2}\int_{\partial\mathcal{M}} d^d x\, g_{0\,ij}\delta(\sqrt{-g_0}\mathcal{T}^{ij}), \end{aligned} \tag{30}$$

where $\mathcal{T}^{ij}$ is the true renormalized stress tensor (of the boundary CFT) and is given by[7]

$$\mathcal{T}_{ij} = \lim_{\epsilon\to 0}\left(\frac{1}{\epsilon^{\frac{d}{2}-1}}T^{ren}_{ij}[\gamma]\right) = \lim_{\epsilon\to 0}\left(-\frac{2}{\sqrt{g(x,\epsilon)}}\frac{\delta S_D^{ren}}{\delta g^{ij}}\right) = -\frac{2}{\sqrt{-g_0}}\frac{\delta S_D^{ren}}{\delta g_0^{ij}}. \tag{31}$$

We can write down the Robin gravity action specific to AdS as

$$S_R^{ren} = S_D^{ren} - \int_{\partial\mathcal{M}} d^d x\, \pi^{ij}_{ren}\gamma_{ij} + \frac{(d-2)}{2}\xi\int_{\partial\mathcal{M}} d^d x\sqrt{-g_0}, \tag{32}$$

which upon variation gives

$$\delta S_R = \text{eq. of motion} + \frac{1}{2}\int_{\partial\mathcal{M}} d^d x\, g_{0\,ij}\;\delta\Big(\sqrt{-g_0}(\mathcal{T}^{ij} + \xi g_0^{ij})\Big), \tag{33}$$

with the variational principle well defined by holding $\sqrt{-g_0}(\mathcal{T}^{ij} + \xi g_0^{ij}) = \mathit{any\ fixed\ quantity}$.

The essential difference between flat space and AdS is that here the variational principle is best formulated in terms quantities that are intrinsic to the field theory: in other words, in terms of a combination of the g_0 and the g_d in the Fefferman-Graham expansion (see [4, 7]) instead of induced metric γ. Note that $\mathcal{T}^{ij}$ is determined in terms of them [7].

Acknowledgments

CK thanks the participants and organizers of the IF-YITP Symposium VI, especially Burin Gumjudpai, Matthew Lake and Shingo Takeuchi for an enjoyable conference, and APCTP for financial support.

[6] We set $D = d + 1$ for convenience.

[7] The boundary we are looking at is time-like $\epsilon = +1$, in the entire AdS discussion.

IOP Conf. Series: Journal of Physics: Conf. Series **883** (2017) 012011 doi:10.1088/1742-6596/883/1/012011

Appendix A. Another Path to Robin?

In the particle mechanics and field theory cases, there exists a direct path from the Dirichlet action to the General Robin action. Let us start with the particle mechanics problem. To the Dirichlet action in (1), add a boundary term $S_B^p = \frac{\xi}{2}\dot{q}^2\big|^T$, the variation of the sum of two gives

$$\delta(S_D^p + S_B^p) = -\int_T dt\,\Big(\ddot{q}^2 + V'(q)\Big)\delta q + \dot{q}\,\delta\Big(\xi\dot{q} + q\Big)\Big|_T. \tag{A.1}$$

This is clearly a well-defined General Robin variational principle. This sort of thing extends trivially to the field theory case as well. Simply consider the following addition to the Dirichlet field theory action:

$$\begin{aligned} S_R' &= \int_{\mathcal{M}} d^D x\sqrt{-g}\left(-\frac{1}{2}g^{\mu\nu}\partial_\mu\phi\partial_\nu\phi - V(\phi)\right) - \frac{\xi}{2}\int_{\partial\mathcal{M}} d^{D-1}x\sqrt{|\gamma|}(\partial_n\phi)^2, &\text{(A.2)}\\ \Rightarrow \delta S_R' &= \int_{\mathcal{M}} d^D x\left(\partial_\mu(\sqrt{-g}g^{\mu\nu}\partial_\nu\phi) - \frac{\partial V(\phi)}{\partial\phi}\right)\delta\phi + \\ &\quad - \int_{\partial\mathcal{M}} d^{D-1}x\sqrt{|\gamma|}\partial_n\phi\,\delta\left(\phi + \xi\partial_n\phi\right) &\text{(A.3)} \end{aligned}$$

For the case of gravity, one might think that an analogous boundary piece can be added to the Dirichlet (Gibbons-Hawking-York) action to produce the General Robin action. This however does not seem to work: for gravity, we find that going through the Neumann action seems to be essential to obtain the Robin action. We describe why this is so, below.

There are two possible terms that could be constructed out of π_{ij}'s that are quadratic, namely, $(\pi^{ij}\gamma_{ij})^2$ and $\pi^{ij}\gamma_{jk}\pi^{kl}\gamma_{li}$. Also, one has to remember that π^{ij} internally contains a $\sqrt{|\gamma|}$ factor, so it would be more advisable to write boundary terms using T^{ij}, the boundary stress tensor[8]. Let us look at the variation of the first candidate, modulo the constants

$$\begin{aligned} \delta S_1 &= \int_{\partial\mathcal{M}} d^{D-1}x\,\delta\big[\sqrt{|\gamma|}\,[\mathrm{Tr}(T)]^2\big] \\ &= \int_{\partial\mathcal{M}} d^{D-1}x\sqrt{|\gamma|}\left(\frac{1}{2}[\mathrm{Tr}(T)]^2\gamma^{ij}\delta\gamma_{ij} + 2\,\mathrm{Tr}(T)\big(T^{ij}\delta\gamma_{ij} + \gamma_{ij}\delta T^{ij}\big)\right). \end{aligned} \tag{A.4}$$

This can be written in terms of π^{ij}'s as

$$\begin{aligned} \delta S_1 &= 4\int_{\partial\mathcal{M}} d^{D-1}x\,\delta\left[\frac{1}{\sqrt{|\gamma|}}(\pi^{ij}\gamma_{ij})^2\right] \\ &= 4\int_{\partial\mathcal{M}} d^{D-1}x\,\frac{1}{\sqrt{|\gamma|}}\left(-\frac{1}{2}(\pi^{ij}\gamma_{ij})^2\gamma^{kl}\delta\gamma_{kl} + 2(\pi^{ij}\gamma_{ij})\pi^{kl}\delta\gamma_{kl} + 2(\pi^{ij}\gamma_{ij})\gamma_{kl}\delta\pi^{kl}\right). \end{aligned} \tag{A.5}$$

The variation of the second candidate term gives,

$$\begin{aligned} \delta S_2 = &= \int_{\partial\mathcal{M}} d^{D-1}x\,\delta\big[\sqrt{|\gamma|}\,\mathrm{Tr}(T^2)\big] \\ &= \int_{\partial\mathcal{M}} d^{D-1}x\sqrt{|\gamma|}\left(\frac{1}{2}\mathrm{Tr}(T^2)\gamma^{ij}\delta\gamma_{ij} + 2\,T^{ij}\delta T_{ij} + 2\,T^{ij}T_j^{\ k}\delta\gamma_{ik}\right). \end{aligned} \tag{A.6}$$

[8] In the following we use the defintions $[\mathrm{Tr}(T)]^2 = (T^{ij}\gamma_{ij})^2$ and $\mathrm{Tr}(T^2) = T^{ij}\gamma_{jk}T^{kl}\gamma_{li}$.

IOP Conf. Series: Journal of Physics: Conf. Series **883** (2017) 012011 doi:10.1088/1742-6596/883/1/012011

This can be written in terms of π^{ij}'s as

$$\begin{aligned}\delta S_2 &= 4\int_{\partial\mathcal{M}} d^{D-1}x\,\delta\left[\frac{1}{\sqrt{|\gamma|}}\left(\pi^{ij}\gamma_{jk}\pi^{kl}\gamma_{li}\right)\right]\\ &= 4\int_{\partial\mathcal{M}} d^{D-1}x\,\frac{1}{\sqrt{|\gamma|}}\Big(-\frac{1}{2}(\pi^{ij}\gamma_{jm}\pi^{mn}\gamma_{ni})\gamma^{kl}\delta\gamma_{kl}\\ &\qquad +2\pi^{ij}\gamma_{jk}\pi^{kl}\delta\gamma_{li}+2\pi^{ij}\gamma_{jk}\gamma_{li}\delta\pi^{kl}\Big).\end{aligned} \tag{A.7}$$

To allow the most general possibility, let us consider adding a combination of these two candidate terms with arbitrary coefficients to the Dirichlet action:

$$S_R' = S_D + \xi\int_{\partial\mathcal{M}} d^{D-1}x\,\frac{1}{\sqrt{|\gamma|}}\,(\pi^{ij}\gamma_{ij})^2 + \zeta\int_{\partial\mathcal{M}} d^{D-1}x\,\frac{1}{\sqrt{|\gamma|}}\,\pi^{ij}\gamma_{jk}\pi^{kl}\gamma_{li}. \tag{A.8}$$

Upon variation this yields

$$\begin{aligned}\delta S_R' &= \text{eom} + \int_{\partial\mathcal{M}} d^{D-1}x\,\pi^{ij}\Big\{\Big[\delta_i^k\delta_j^l + \frac{\xi}{\sqrt{|\gamma|}}\Big(-\frac{1}{2}\pi^{mn}\gamma_{mn}\gamma_{ij}\gamma^{kl}+2\pi^{mn}\gamma_{mn}\delta_i^k\delta_j^l\Big)\\ &\qquad +\frac{\zeta}{\sqrt{|\gamma|}}\Big(-\frac{1}{2}\pi^{mn}\gamma_{im}\gamma_{jn}\gamma^{kl}+2\pi^{nl}\gamma_{jn}\delta_i^k\Big)\Big]\delta\gamma_{kl}\\ &\qquad +\frac{2}{\sqrt{|\gamma|}}\Big[\xi\gamma_{ij}\gamma_{kl}+\zeta\gamma_{ik}\gamma_{jl}\Big]\delta\pi^{kl}\Big\}\end{aligned} \tag{A.9}$$

For this to reduce to a General Robin variation, we need the coefficient of $\delta\pi^{kl}$ to be some number times the coefficient of $\delta\gamma_{kl}$. This is clearly impossible for any choice of ζ and ξ.

The essential difference between mechanics/field theory and gravity is that here, the $\sqrt{|\gamma|}$ term shows up, which is essentially non-polynomial. This makes the Neumann term an essential intermediate step in our path to Robin: there does not seem to be direct path to it from the Dirichlet (Gibbons-Hawking) boundary term.

Another way to state the same observation is that one can view both Dirichlet and Neumann boundary conditions as limits of Robin in mechanics and field theory, but in general relativity only the Neumann boundary condition can be viewed as a limit of Robin. At the technical level, the problem is that for γ, the key relation (25) holds, but for π there is no such relation.

References

[1] G. W. Gibbons and S. W. Hawking "Action integrals and partition functions in quantum gravity," Phys. Rev. D **15**, 2752 (1977);
J. W. York,Jr., "Role of Conformal Three-Geometry in the Dynamics of Gravitation," Phys. Rev. Lett. **28**, 1082 (1972).

[2] C. Krishnan and A. Raju, "A Neumann Boundary Term for Gravity," arXiv:1605.01603 [hep-th].

[3] C. Krishnan, K. V. P. Kumar and A. Raju, "An alternative path integral for quantum gravity," JHEP **1610**, 043 (2016) [arXiv:1609.04719 [hep-th]].

[4] C. Krishnan, A. Raju and P. N. B. Subramanian, "A Dynamical Boundary for Anti-de Sitter Space," Phys. Rev. D **94**, no. 12, 126011 (2016), arXiv:1609.06300 [hep-th].

[5] J. D. Brown and J. W. York, Jr., "Quasilocal energy and conserved charges derived from the gravitational action," Phys. Rev. D **47**, 1407 (1993) [gr-qc/9209012].

[6] C. Krishnan, "Quantum Field Theory, Black Holes and Holography," arXiv:1011.5875 [hep-th].

[7] S. de Haro, S. N. Solodukhin and K. Skenderis, "Holographic reconstruction of space-time and renormalization in the AdS / CFT correspondence," Commun. Math. Phys. **217**, 595 (2001) [hep-th/0002230].

IOP Conf. Series: Journal of Physics: Conf. Series **883** (2017) 012012 doi:10.1088/1742-6596/883/1/012012

Decoherence of Topological Qubit in Linear Motions: Decoherence Impedance, Anti-Unruh and Information Backflow

Pei-Hua Liu and Feng-Li Lin

Department of Physics, National Taiwan Normal University, Taipei, 11677, Taiwan

E-mail: `lphhpl@gmail.com`, `fengli.lin@gmail.com`

Abstract. In this work we study the decoherence of topological qubits in linear motions. The topological qubit is made of two spatially-separated Majorana zero modes which are the edge excitations of Kitaev chain [1]. In a previous work [2], it was shown by one of us and his collaborators that the decoherence of topological qubit is exactly solvable, moreover, topological qubit is robust against decoherence in the super-Ohmic environments. We extend the setup of [2] to consider the effect of motions on the decoherence of the topological qubits. Our results show the thermalization as expected by Unruh effect. Besides, we also find the so-called "anti-Unruh" phenomena which shows the rate of decoherence is anti-correlated with the acceleration in short-time scale. Moreover, we modulate the motion patterns of each Majorana modes and find information backflow and the preservation of coherence even with nonzero accelerations. This is the characteristics of the underlying non-Markovian reduced dynamics. We conclude that he topological qubit is in general more robust against decoherence than the usual qubits, and can be take into serious consideration for realistic implementation to have robust quantum computation and communication. This talk is based on our work in [3].

1. Introduction

In the past twenty years, quantum information sciences have achieved tremendous progress both theoretically and experimentally. However, it is still far from reaching out a large-scale quantum computer. The main obstacle is short of realistic robust qubits against quantum decoherence so that it is hard to scale up the quantum resource for computation. One way to break this obstacle is to find the qubits which are topological in nature ensured by the underlying quantum dynamics. One candidate for such a qubit is the edge excitations of the topological insulators and superconductors, for which their robustness is guaranteed dynamically. In [2] we have demonstrated that the topological qubit made of two Majorana zero modes of the Kitaev chain [1] is indeed robust against decoherence in the super-Ohmic environment. The interaction Hamiltonian between the Majorana zero modes and the environment is given by

$$V := \gamma_1 O_1 + \gamma_2 O_2 \tag{1}$$

where $\gamma_{1,2}$ are the Majorana zero modes, i.e., they satisfy the Clifford algebra: $\gamma_1^2 = \gamma_2^2 = 1$ and $\gamma_1\gamma_2 + \gamma_2\gamma_1 = 0$. The operators O_i's are the anti-Hermitian environmental operators obeying the locality constraint

$$\langle \mathcal{O}_1(t)\mathcal{O}_2(t')\rangle = 0 \tag{2}$$

IOP Conf. Series: Journal of Physics: Conf. Series **883** (2017) 012012 doi:10.1088/1742-6596/883/1/012012

where $\mathcal{O}_i(t) := e^{iH_Et}O_ie^{-iH_Et}$ with H_E the environmental Hamiltonian.

Moreover, it was shown in [2] that the open-system quantum reduced dynamics of the topological qubit is exactly solvable, and the result is as follows. Given the initial state of the topological qubit

$$\rho_M(t=0) = \begin{pmatrix} \rho_{00} & \rho_{01} \\ \rho_{01}^* & 1-\rho_{00} \end{pmatrix}, \tag{3}$$

its reduced dynamics has been solved in [2] and takes the simple closed form

$$\rho_M(t) = \begin{pmatrix} \frac{1}{2}+(\rho_{00}-\frac{1}{2})e^{\mathcal{I}_1(t)+\mathcal{I}_2(t)} & e^{\mathcal{I}_2(t)}\text{Re }\rho_{01}+ie^{\mathcal{I}_1(t)}\text{Im }\rho_{01} \\ e^{\mathcal{I}_2(t)}\text{Re }\rho_{01}-ie^{\mathcal{I}_1(t)}\text{Im }\rho_{01} & \frac{1}{2}-(\rho_{00}-\frac{1}{2})e^{\mathcal{I}_1(t)+\mathcal{I}_2(t)} \end{pmatrix} \tag{4}$$

where the "influence functional" associated with the i-th Majorana zero mode is

$$\mathcal{I}_i(t) := 2\int^t d\tau \int^t d\tau'\, \overline{G}_{i,sym}(\tau-\tau') \tag{5}$$

with

$$\overline{G}_{i,sym}(t-t') = \frac{1}{2}(\langle\mathcal{O}_i(t)\mathcal{O}_i(t')\rangle + \langle\mathcal{O}_i(t')\mathcal{O}_i(t)\rangle)\,. \tag{6}$$

Here, $\overline{G}_{i,sym}$ is the so-called Majorana-dressed symmetric Green function, which is different from the usual fermionic one, i.e., $G_{i,sym}(t-t') = \frac{1}{2}\left(\langle\mathcal{O}_i(t)\mathcal{O}_i(t')\rangle - \langle\mathcal{O}_i(t')\mathcal{O}_i(t)\rangle\right)$. The sign change is due to the Majorana dressing so that time-ordering for $\mathcal{O}_i$ changes from the fermionic one to the bosonic one. Based on this reduced density matrix, it is straightforward to see that the quantum dynamics is robust for the super-Ohmic environments, as shown in [2].

2. Topological qubits in linear motions

We consider the decoherence patterns of a topological qubit made of two Majorana zero modes in the generic linear and circular motions in the Minkowski spacetime. Based on the formulation of the topological qubit and its reduced dynamics developed in [2], we develop the formalism to study the reduced dynamics for a moving topological qubit by generalizing the one given in [4] for the usual Unruh-DeWitt (UDW) detector. We show that the reduced dynamics is exact without Markov approximation. We can also obtain the exact transition probability of the topological qubit beyond the Markovian approximation.

The key idea of [4] is that the motion of the particle is induced by performing proper coordinate transformations on the evolving Hamiltonians for either UDW detector or the environmental fields. In this work we consider the linear motions of the Majorana zero modes as shown in Fig. 1. As shown in (4), the reduced dynamics is encoded in the influence functionals. In the case of linear motion, we can choose to observe the reduced dynamics in the comoving frame of k-th Majorana zero modes (called this frame as "M_k"-frame), then the influence functional associated with the i-th Majorana zero modes as function of proper time τ_k is

$$\begin{aligned} \mathcal{I}_i^{M_k}(\tau_k) &= -2\int_0^{\tau_k} d\tau_k' \int_0^{\tau_k} d\tau_k''\, \lambda_i(\tau_i(\tau_k'))\lambda_i(\tau_i(\tau_k''))\int_{-\infty}^{\infty} d\omega\, |\omega|^{d-1}\mathcal{A}_i(\omega)e^{-i\omega(t(\tau_k')-t(\tau_k''))} \\ &\quad\times \oint d\mathbf{\Omega}\exp\left[i\omega\hat{n}\cdot(\vec{x}(\tau_k')-\vec{x}(\tau_k''))\right] \end{aligned} \tag{7}$$

where $\hat{n} := \vec{k}/|\vec{k}|$ is the unit normal to the unit sphere of measure $\oint d\mathbf{\Omega}$, d is the space dimension, and $\mathcal{A}_i(\omega)$ is the environmental spectral density. On the other hand, we can choose to observe the

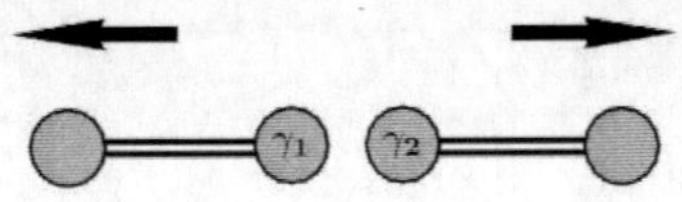

Figure 1. A pair of topological qubits with their constituent Majorana zero modes in incoherent linear boost or acceleration. The Majorana zero modes $\gamma_{1,2}$ form a topological qubit of which we study the effect of the motions to the decoherence pattern.

reduced dynamics in the environmental frame (called "E-frame"), then the influence functional associated with i-th Majorana zero mode becomes

$$\begin{aligned} \mathcal{I}_i^E(t) &= -2\int_0^t dt' \int_0^t dt''\, \lambda_i(\tau_i(t'))\lambda_i(\tau_i(t''))\frac{d\tau_i}{dt'}\frac{d\tau_i}{dt''}\int_{-\infty}^{\infty} d\omega\, |\omega|^{d-1}\mathcal{A}_i(\omega)e^{-i\omega(t'-t'')} \\ &\quad\times \oint d\mathbf{\Omega}\exp\left[i\omega\hat{n}\cdot[\vec{x}_i(\tau_i(t'))-\vec{x}_i(\tau_i(t''))]\right]. \end{aligned} \tag{8}$$

In both case, the worldline relations $t = t(\tau_i)$ and $\vec{x} = \vec{x}(\tau_i)$ are implicitly used.

Moreover, we can extract from the reduced dynamics the transition probability of UDW detector as follows:

$$P_{i\to f} = \lim_{\mathcal{T}\to\infty}\langle f|\sum_{\mathbf{m}}\langle\mathbf{m}|U(t)|\mathbf{0}\rangle|i\rangle\langle i|\langle\mathbf{0}|U^\dagger(t)|\mathbf{m}\rangle|f\rangle = \lim_{\mathcal{T}\to\infty}\langle f|\,\mathrm{Tr}_E\,\rho^D(t)|f\rangle \tag{9}$$

where $\mathrm{Tr}_E\,\rho^D(t)$ is the reduced density matrix for the (M)UDW detector and note that $\rho^D(t=0) = |\mathbf{0}\rangle|i\rangle\langle i|\langle\mathbf{0}|$ so that $\rho^D(t) = U(t)\rho^D(t=0)U^\dagger(t)$. Note that unlike the usual treatment for the UDW detector, the transition probability derived here is exact, not just up to first order of coupling between probe and environment. We can then compare our results of Unruh effect with the ones obtained by usual treatment.

3. Summary of our results

Despite that the topological qubit is novel and non-local, the effect of the (acceleration) motions to its decoherence patterns may bear some generic features shared by the usual qubit in the similar motions. In some sense, it can be thought as the manifestation of the Unruh effect in the decoherence. Here we list what we found in this work, which we think should be generic:

- Thermalization: We find that the acceleration does cause thermalization as expected by the Unruh effect. This can be seen by the complete decoherence of the accelerating topological qubit in the super-Ohmic environment as this kind of qubits without acceleration are robust against decoherence. However, the influence functional is different from the thermal one. It is worthy of further study to see if the influence functional of the usual qubit in the constant acceleration is the same as the thermal one or not.
- Decoherence impedance: By studying the decoherence patterns we find some novel feature. it seems that the topological qubit is resisting the enforced decoherence due to the sudden change of the motion status such as acceleration or boost. We called this tendency the

IOP Conf. Series: Journal of Physics: Conf. Series **883** (2017) 012012 doi:10.1088/1742-6596/883/1/012012

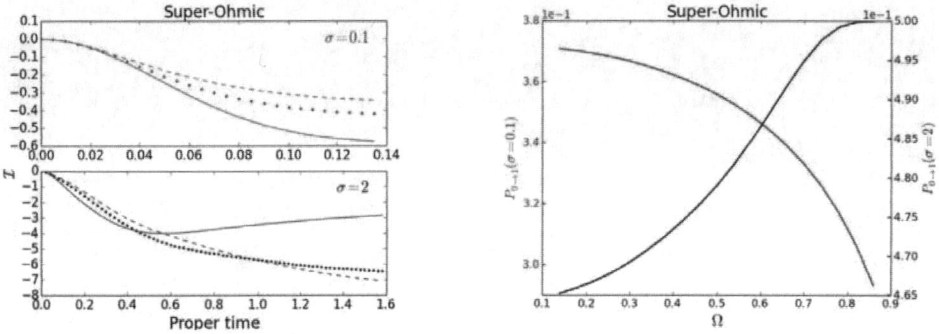

Figure 2. Decoherence patterns and transition probability $P_{0\to1}$ in the M-frame of a circularly moving MUDW detector of constant angular velocity Ω in the super-Ohmic environments with the switching function of time duration scales: $\sigma = 0.1$ (red) and $\sigma = 2$ (blue). Left : Decoherence patterns for $\Omega = 0.7$ (solid), $\Omega = 0.9$ (dashed) and $\Omega = 0.95$ (dotted). Right: $P_{0\to1}$ versus Ω. This figure shows that the decoherence impedance and "anti-Unruh" imply each other.

decoherence impedance, namely, the initial rate of decoherence is anti-correlated with the size of change, such as acceleration. In the late time, this impedance effect will be taken over by thermalization so that in the mean time the overtaking of the decoherence rates happens. As a simple explicit example, it is shown in the left panel of Fig.2 for circularly moving detector.

- Anti-Unruh: As shown in (10), in our formalism we can obtain the full transition probability of the topological qubit as the UDW detector, which can be used to characterize Unruh effect.

$$P_{0\to1} = \lim_{t\to\infty} \frac{1}{2}(1 - e^{\mathcal{I}_1(t)+\mathcal{I}_2(t)}) \tag{10}$$

  where $\mathcal{I}_i(t)$ is the influence functional.
  Follow the scheme in [5] by decoupling the UDW detector from the environmental field after finite time duration, we can find from the transition probability the aforementioned "anti-Unruh" phenomenon for the accelerating topological qubit in both linear and circular motions. Our result is exact without Markov approximation as done in [5] and thus justifies the "anti-Unruh" phenomenon is generic. An explicit example is shown in the right panel of Fig. 2 for circularly moving detector.
  In fact, both the decoherence patterns ($\propto e^{\mathcal{I}_i(t)}$) and the transition probability are all dictated by the influence functional. We would expect the decoherence impedance and "anti-Unruh" phenomenon imply each other as both are the short-time non-equilibrium effects. This is indeed what we observe in this work.
- Information backflow and time modulation: As our reduced dynamics is exact without Markov approximation, we shall expect some non-Markovian behaviors characterized by the information flow. We thus apply the time modulations to the coupling constant and the acceleration to see if some modulations can invoke information backflow. We find that not all modulations will bring out information backflow but some does and sometimes even ensures the robustness against decoherence. We have done for the amplitude modulation

$$a(\tau) = \begin{cases} C, & \text{if } \tau_1 < \tau < \tau_2, \\ -C, & \text{if } \tau_2 < \tau < 2\tau_2 - \tau_1, \\ 0, & \text{otherwise.} \end{cases} \tag{11}$$

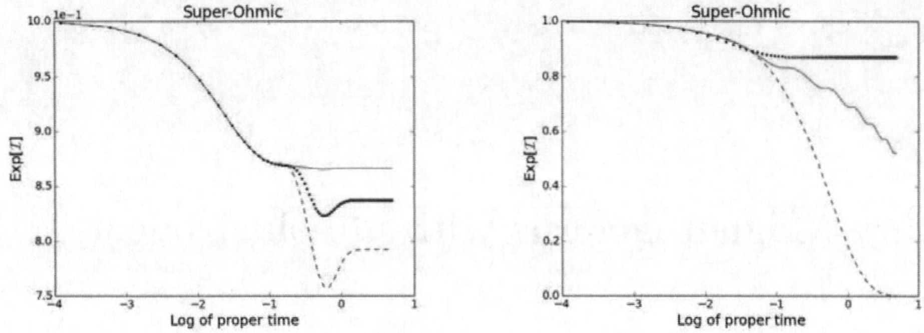

Figure 3. Decoherence patterns in the M-frame of an accelerating topological qubit in the super-Ohmic environment with the amplitude modulation (AM), i.e., Eq. (11) and frequency modulation (FM), i.e., Eq. (12) of the acceleration. Left: AM with $C = 1$ (solid red) , $C = 5$ (dotted blue) and $C = 10$ (dashed black) as $\tau_1 = 0.3$ and $\tau_2 = 0.5$. Right: FM with $\omega_G = 10$ (solid red) , $\omega_G = 50$ (dotted blue) and $\omega_G = 1$ (dashed black) as $a = 10$.

and the frequency modulation

$$a(\tau) = a \cos \omega_G \tau \ . \tag{12}$$

The results are shown in Fig. 3.

Moreover, by really exploiting the non-local feature of the topological qubit, we consider the frame dependence issues when both Majorana zero modes move incoherently. We indeed find some novel feature: some incoherent accelerations will instead preserve the coherence of the topological qubit so that it will not be thermalized by the accelerations. This is a phenomenon peculiar for the topological qubit as its constituents can move differently.

The exact solvability of the reduced dynamics even with the nontrivial motions of the topological qubit and the results presented in this paper demonstrate the power of marrying topological order with relativistic quantum information. It is interesting to see if some of the "generic" features aforementioned will also occur for the usual qubit. The answer to this question will definitely shed new light on our understanding of relativistic quantum information.

Acknowledgements

FLL would like to thank APCTP for the travel support to attend and present this work in IF-YITP GR+HEP+Cosmo International Symposium VI held at IF, Naresuan University, Thailand, 3-5 August, 2016. This work is supported by Taiwan Ministry of Science and Technology through Grant No. 103-2112-M-003 -001 -MY3 and 103-2811-M-003 -024.

References

[1] A. Y. Kitaev, "Unpaired Majorana fermions in quantum wires", Phys. Usp. **44**, 131 (2001). [arXiv:cond-mat/0010440].
[2] Shih-Hao Ho, Sung-Po Chao ,Chung-Hsien Chou and Feng-Li Lin,"Decoherence patterns of topological qubits from Majorana modes," New J. Phys. **16**, no. 11, 113062 (2014). [arXiv:1406.6249 [cond-mat.str-el]].
[3] P. H. Liu and F. L. Lin, "Decoherence of Topological Qubit in Linear and Circular Motions: Decoherence Impedance, Anti-Unruh and Information Backflow," JHEP **1607**, 084 (2016) [arXiv:1603.05136 [quant-ph]].
[4] E. G. Brown, E. Martin-Martinez, N. C. Menicucci and R. B. Mann, "Detectors for probing relativistic quantum physics beyond perturbation theory," Phys. Rev. D **87**, 084062 (2013). [arXiv:1212.1973 [quant-ph]].
[5] W. G. Brenna, R. B. Mann and E. Martin-Martinez, "Anti-Unruh Phenomena," arXiv:1504.02468 [quant-ph].

IOP Conf. Series: Journal of Physics: Conf. Series **883** (2017) 012013 doi:10.1088/1742-6596/883/1/012013

Inflationary Magnetogenesis with On-shell Local $U(1)$ Symmetry

Guillem Domènech, Chunshan Lin and Misao Sasaki

Center for Gravitational Physics, Yukawa Institute for Theoretical Physics, Kyoto University

Abstract. We propose a new mechanism for inflationary magnetogenesis in which the local $U(1)$ symmetry is broken during inflation. Nevertheless it is shown that the $U(1)$ symmetry is recovered on-shell. We find that it is free from both the strong coupling and back reaction problems, and can explain the origin of cosmic magnetic fields on intergalactic scales, whose existence has been strongly suggested by recent observations.

1. Introduction

Magnetic fields are present throughout the universe and play an important role in many astrophysical process. However, their origin has not been well understood yet. Recently it was reported that large scale magnetic fields of $\gtrsim 10^{-15}$G with coherence length of $\gtrsim$ Mpc may be present in inter-galactic space [1, 2, 3, 4]. If such a magnetic field exists, no causal process, not even dynamo, after the recombination time ($z \sim 1000$) can explain its origin. And it becomes even more difficult to explain with processes at $z > 1000$ because of the horizon problem.

It is then natural to attribute the origin of such a large scale magnetic field to inflation. However, because the Friedmann universe is conformally flat, the conformal invariance of the electromagnetic field makes it impossible to allow any process that amplifies the magnetic field. The electromagnetic field remains in vacuum if it is in vacuum in the beginning, while any excitation from vacuum would have been diluted exponentially during inflation.

Many models with broken conformal invariance have been proposed [5, 6] One of the simplest, $U(1)$ invariant models is [7, 8]

$$\mathcal{L}_{EM} = -\frac{1}{4} f^2(\phi) F_{\mu\nu} F^{\mu\nu}, \tag{1}$$

where $F_{\mu\nu} \equiv \partial_\mu A_\nu - \partial_\nu A_\mu$. This model is equivalent to Bekenstein's variable charge theory [9].

Unfortunately, it was shown that the model suffers from either back-reaction or strong coupling problem [10]. If f grows in time, the effective coupling becomes too large in the past, while if f decays in time, the energy density of the electromagnetic field becomes too large to be regarded as a small perturbation to inflation.

IOP Conf. Series: Journal of Physics: Conf. Series **883** (2017) 012013 doi:10.1088/1742-6596/883/1/012013

2. On-shell $U(1)$ symmetric model

To circumvent the above mentioned difficulty, we propose a model in which f grows in time, but it is free from both strong coupling and back-reaction problems. We consider the Lagrangian,

$$\begin{aligned}\mathcal{L} = \quad & -\frac{1}{4}f^2(\phi)F_{\mu\nu}F^{\mu\nu} - e_0 f(\phi)\gamma^\mu A_\mu \bar{\psi}\psi \\ & + i\bar{\psi}\gamma^\mu\left(\partial_\mu + \Gamma_\mu\right)\psi - m\bar{\psi}\psi, \end{aligned} \tag{2}$$

where e_0 is the value of the electric charge today, ψ is a charged fermion and and Γ_μ is the spin connection. For slowly varying f, e.g., $\dot{f}/f = O(H)$ where H is the Hubble parameter, we may absorb f into A_μ by scaling $A_\mu \to f^{-1}A_\mu$. This gives

$$\begin{aligned}\mathcal{L} = \quad & -\frac{1}{4}F_{\mu\nu}F^{\mu\nu} + i\bar{\psi}\gamma^\mu D_\mu\psi - m\bar{\psi}\psi \\ & -F_{\mu\nu}Q^\mu A^\nu - \frac{1}{2}Q_\mu Q^\mu A_\nu A^\nu + \frac{1}{2}\left(Q_\mu A^\mu\right)^2, \end{aligned} \tag{3}$$

where $Q_\mu \equiv -\partial_\mu(\ln f)$ and $D_\mu \equiv \partial_\mu + ie_0 A_\mu + \Gamma_\mu$. From this it is obvious that the model is free from the strong coupling problem even if f grows in time.

Of course the price to pay is the broken local $U(1)$ gauge symmetry. Nevertheless, if we use the equations of motion,

$$\nabla_\mu(f^2F^{\mu\nu}) - e_0 f\bar{\psi}\gamma^\nu\psi = 0, \tag{4}$$

the divergence of it gives

$$\nabla_\mu\left(e_0 f\bar{\psi}\gamma^\mu\psi\right) = 0, \tag{5}$$

which implies

$$\begin{aligned}\delta_g S &= -\int\sqrt{-g}\,e_0 f\partial_\mu\sigma\bar{\psi}\gamma^\mu\psi \\ &= -\int\sqrt{-g}\left[\nabla_\mu\left(e_0 f\sigma\bar{\psi}\gamma^\mu\psi\right) - \sigma\nabla_\mu\left(e_0 f\bar{\psi}\gamma^\mu\psi\right)\right], \end{aligned} \tag{6}$$

where $\delta_g A_\mu = \partial_\mu\sigma$. Thus the $U(1)$ gauge symmetry is recovered on-shell. This suggests that this $U(1)$ symmetry breaking may not be so harmful after all.

3. Magnetogenesis

It is known that if the function f behaves as $f^2 = a^4$, one would obtain a scale-invariant spectrum for the B field [11, 10]. But this implies the logarithmic IR divergence in the energy density. To avoid this, we assume

$$f^2(N) = \exp[-4N + (4\epsilon + \nu)N]\,; \quad N > 0\,, \tag{7}$$

where $\epsilon \equiv -\dot{H}/H^2$ is the slow-roll parameter during inflation, ν is a new parameter which is required to satisfy $0 < \nu \ll 1$, and N is the number of e-folds counted backward from the end of inflation. The requirement $0 < \nu \ll 1$ guarantees the finiteness of the total energy density and an almost scale-invariant B spectrum. Note that the above behavior of f may be fairly easily realized, e.g. in power-law inflation, by

$$f = f_0 \exp\left[\alpha\frac{\phi}{M_p}\right] \tag{8}$$

with $\alpha = O(1)$, and $M_p = (8\pi G)^{-1/2}$ is the Planck mass.

Because the model is essentially the same as the one discussed, e.g. in [11], apart from the $U(1)$ breaking interaction with the charged fermion, one can simply repeat the computation given in the literature. The resulting magnetic spectrum is

$$P_B(k; N) = \frac{H(N)^4}{2\pi^2} e^{\nu(N-N_k)}, \tag{9}$$

where N_k is the number of e-folds at which the comoving scale k crossed the horizon. The energy density of the magnetic field is given by integrating over the whole momentum space (with UV cutoff at $k = aH \equiv k_*(N)$),

$$\rho_B(N) = \frac{1}{2}\int_0^{k_*(N)} \frac{dk}{k} P_B = \frac{H^4(N)}{4\pi^2\nu}. \tag{10}$$

Now we require that the model is free from backreaction. Namely ρ_B must be much smaller than background energy density. This means

$$\frac{H_f^2}{M_p^2} \ll \nu \ll 1. \tag{11}$$

For reasonable choices of H_f this is easily satisfied.

The magnetic field at the end of inflation is

$$B_f \sim H_f^2 \nu^{-1/2} = 10^{-12}\nu^{-1/2}\left(\frac{H_f}{10^{-6}M_p}\right)^2 M_p^2. \tag{12}$$

Assuming an instantaneous reheating, the current amplitude of the B field is estimated as

$$B_0 \sim B_f \times \frac{T_{\rm CMB}^2}{M_p H_f} \sim 10^{-12}\nu^{-1/2}\left(\frac{H_f}{10^{-6}M_p}\right) G, \tag{13}$$

where T_{CMB} is the current CMB temperature. This satisfies the most recent observational constraint $B_{\rm Mpc} < 10^{-9}G$ by Planck [12], for

$$10^{-6} < \nu \ll 1. \tag{14}$$

4. Conclusion

We proposed a new model for inflationary magnetogenesis in which the local $U(1)$ symmetry is broken during inflation. Interestingly, in our model the broken $U(1)$ symmetry was shown to be recovered on-shell. We found that the model is free from both the strong coupling and back reaction problems, and can explain the origin of intergalactic magnetic fields whose existence was strongly suggested by recent observations.

Acknowledgments

MS would like to thank the local organizers of the IF-YITP symposium VI at Naresuan University, August 2016, for their warm hospitality during the conference. This work was supported in part by MEXT KAKENHI Nos. 15H05888 and 15K21733, and by JSPS KAKENHI No. 15F15321. This talk is based on [13].

IOP Conf. Series: Journal of Physics: Conf. Series **883** (2017) 012013 doi:10.1088/1742-6596/883/1/012013

References

[1] A. Neronov, I. Vovk, Science 328:73-75,2010, [arxiv: 1006.3504].
[2] F. Tavecchio, G. Ghisellini, G. Bonnoli and L. Foschini, MNRAS (2011) 414 (4): 3566-3576.
[3] F. Tavecchio, G. Ghisellini, L. Foschini, G. Bonnoli, G. Ghirlanda and P. Coppi, MNRAS (2010) 406 (1): L70-L74.
[4] A. M. Taylor, I. Vovk, A. Neronov, Astronomy & Astrophysics, Volume 529, 2011.
[5] A. Kandus, K. E. Kunze and C. G. Tsagas, Phys. Rept. **505**, 1 (2011).
[6] R. Durrer and A. Neronov, Astron. Astrophys. Rev. **21**, 62 (2013).
[7] M. S. Turner and L. M. Widrow, Phys. Rev. D 37 (1988) 2743.
[8] B. Ratra, Astrophys. J. 391, L1 (1992).
[9] J.D. Bekenstein. Phys. Rev. D25 (1982) 1527
[10] V. Demozzi, V. Mukhanov and H. Rubinstein, JCAP **0908**, 025 (2009).
[11] K. Bamba and M. Sasaki, JCAP **0702**, 030 (2007)
[12] P. A. R. Ade *et al.* [Planck Collaboration], arXiv:1502.01594 [astro-ph.CO].
[13] G. Domenech, C. Lin and M. Sasaki, Europhys. Lett. **115**, no. 1, 19001 (2016).

Schwinger effect for non-Abelian gauge bosons

Michael Ragsdale

E-mail: raggy65@mail.fresnostate.edu
Department of Physics, California State University Fresno, Fresno, CA 93740-8031, USA

Douglas Singleton

E-mail: dougs@csufresno.edu
Department of Physics, California State University Fresno, Fresno, CA 93740-8031, USA
and
Institute of Experimental and Theoretical Physics Al-Farabi KazNU, Almaty, 050040, Kazakhstan

Abstract. We investigate the Schwinger effect for the gauge bosons in an unbroken non-Abelian gauge theory (*e.g.* the gluons of QCD). We consider both constant "color electric" fields and "color magnetic" fields as backgrounds. As in the Abelian Schwinger effect we find there is production of "gluons" for the color electric field, but no particle production for the color magnetic field case. Since the non-Abelian gauge bosons are massless there is no exponential suppression of particle production due to the mass of the electron/positron that one finds in the Abelian Schwinger effect. Despite the lack of an exponential suppression of the gluon production rate due to the masslessness of the gluons, we find that the critical field strength is even larger in the non-Abelian case as compared to the Abelian case. This is the result of the confinement phenomenon on QCD.

1. Introduction

The Schwinger effect [1, 2] is the creation of electron-positron pairs from a uniform electric field. The energy to create the pairs comes from energy of the electric field. Since the rest mass energy of the electron is large relative to the electric field energy density that one can achieve in the laboratory the Schwinger effect has not been observed experimentally in the form in which it was first calculated – a uniform background electric field producing electron-positron pairs. The reason for this is the probability per unit volume per unit time of creating e^+e^- pairs is given by

$$\frac{\text{prob}_{e^+e^-}}{\text{Vol} \times \text{time}} \propto e^2 E_{EM}^2 \exp\left(-\frac{\pi m^2}{e E_{EM}}\right) , \tag{1}$$

where e and m are the charge and mass of the electron/positron, E_{EM} is the magnitude of the uniform electric field and c and $\hbar$ have been set to 1. Since the electron/positron have a finite rest mass the exponential will suppress e^+e^- production unless $\frac{\pi m^2}{eE_{EM}} \sim 1$. If one takes $\frac{\pi m^2}{eE_{EM}} = 1$, restores factors of c and $\hbar$ then one finds the this uniform electric field magnitude is enormous – $E_{EM} = 1.4 \times 10^{14} \frac{dyne}{esu}$ or $E_{EM} = 4.2 \times 10^{18} \frac{N}{C}$ – well beyond the present ability to create in a laboratory. If electron/positrons were lighter, or massless, one would more readily be able to observe the electromagnetic Schwinger effect.

While the electron/positron mass is non-zero there is a system where a charged particle is massless and thus the Schwinger effect should not have the exponential suppression found in the electromagnetic case. The is the case of gluons in quantum chromodynamics (QCD). Gluons are massless and carry color charge due to the non-Abelian nature of QCD. Thus we want to investigate the QCD version of the Schwinger effect whereby a constant background color "electric" field creates gluons. Of course due to color confinement one can not really make a constant color field over a macroscopic distance. However within QCD bound systems one might think of the quantum chromodynamic flux tubes that are postulated to bind quarks together into mesons/baryons as giving a uniform color electric and color magnetic fields inside the tubes.

In this work we consider, for simplicity, the $SU(2)$ non-Abelian gauge theory. This has almost all of the features of larger non-Abelian groups like $SU(3)$ but the details of the calculations are simpler and more transparent. Often in this work we will mention "gluons" or QCD which technically refer to the strong $SU(3)$ gauge group, but we mean by this the $SU(2)$ toy model of the true $SU(3)$ interaction.

2. Brief review of electric Schwinger effect

In this section we will give a very brief overview of Schwinger's method for calculating pair production in the case of Abelian electromagnetic fields. In the following section we will use this background to address pair production in a non-Abelian color electric and color magnetic fields. In addition to the original articles on the Schwinger effect [1, 2] there are many good discussion of this method in the literature. A very small sample of these are found in references [3] [4] [5] [6]. We will follow most closely the pedagogical article [3].

To begin the vacuum to vacuum transition amplitude is given very generally by the expression

$$\text{amp}(\text{vac} \to \text{vac}) \longrightarrow \int [d\phi] \exp\left(i \int d^4x \mathcal{L}(\phi, \partial_\mu \phi) \right) , \tag{2}$$

where ϕ is some generic field, $\int [d\phi]$ is a path integral over different field configurations, and $\mathcal{L}(\phi, \partial_\mu \phi)$ is the Lagrange density for the field ϕ. The Lagrange density will lead to an equation of motion for the field $\phi(x)$ in terms of some operator $\mathcal{O}\phi$ [1]. The field $\phi(x)$ can be expanded as $\phi(x) = \sum_n a_n \chi_n(x)$ where $\chi_n(x)$ are eigenstates of the equation of motion given by

$$\mathcal{O}\chi_n(x) = \lambda_n \chi_n(x) , \tag{3}$$

with λ_n being the eigenvalues. In terms of the operator $\mathcal{O}$ the vacuum to vacuum amplitude is $\text{amp}(\text{vac} \to \text{vac}) \sim \frac{const.}{det\mathcal{O}}$ where the determinant of the operator can be written in terms the eigenvalues as $det\mathcal{O} = \prod_n \lambda_n$. Using all this we can write out the results as

$$\frac{1}{det\mathcal{O}} = \exp[-\ln(det\mathcal{O})] = \exp\left[-\ln \prod_n \lambda_n \right] = \exp\left[-\sum_n \ln \lambda_n \right] = \exp[-tr \ln(\mathcal{O})] . \tag{4}$$

Next we use the representation of the logarithm as $\ln \lambda_n = -\int_0^\infty \frac{\exp(-\lambda_n s)}{s} ds$ to write the middle expression in (4) as

$$tr(\ln \mathcal{O}) = -\sum_n \int_0^\infty \frac{\exp(-\lambda_n s)}{s} ds \equiv \zeta . \tag{5}$$

[1] In the example used in Holstein [3] of a scalar field of mass m and charge e the operator is $\mathcal{O} = (\partial_\mu + ieA_\mu(x))^2 + m^2$ where A_μ is the electromagnetic vector potential of the electromagnetic field.

IOP Conf. Series: Journal of Physics: Conf. Series **883** (2017) 012014 doi:10.1088/1742-6596/883/1/012014

Finally the vacuum to vacuum amplitude from (2) becomes

$$\text{amp}(\text{vac} \to \text{vac}) \propto \exp[-\zeta] \ . \tag{6}$$

If ζ has a real part (*i.e.* $\gamma = Re(\zeta)$) then one can square the amplitude to get the probability for pair production as

$$\text{prob}_{pair} = (1 - \exp[-2\gamma]) \ . \tag{7}$$

This is the basic procedure which we will apply to the pair production of gluons in uniform color electric and color magnetic fields. However before moving on to this we give a few more details about the pair production of scalar particles of mass m and charge e in a uniform electric field. For a uniform electric field in the z-direction $\mathbf{E} = E_0\hat{z}$ the vector potential can be of the form $\mathbf{A}(t) = -E_0 t\hat{z}$ or $\phi(z) = -E_0 z$. (In the QCD case we will find a similar situation for the QCD potentials). For the time-dependent vector potential, $\mathbf{A}(t) = -E_0 t\hat{z}$, the operator in footnote 1 becomes

$$\mathcal{O} = \partial_t^2 - (\partial_z - ieE_0 t)^2 - \partial_x^2 - \partial_y^2 + m^2 \ . \tag{8}$$

The eigenvalues connected with (8) are

$$\lambda_n = eE_0(2n+1) + p_x^2 + p_y^2 + m^2 \ , \qquad \text{n} = 0, 1, 2... \ . \tag{9}$$

Using this eigenvalues in (5) and performing the sum [2] of the different pieces of λ_n yields

$$\zeta = -iL^3 T \frac{eE_{EM}}{16\pi^2} \int_0^\infty \frac{ds}{s^2} \frac{\exp[-m^2 s]}{\sin(eE_{EM}s)} \ , \tag{10}$$

where L and T are the spatial and temporal size to "cube" inside which the system is quantized. The factor of i comes from doing a rotation to imaginary time and back (*i.e.* $t \to iT$). This time rotation also involves the change $E_{EM} \to iE_{EM}$. We will use the same procedure when we carry out the QCD version of this calculation. In order to have particle production ζ needs to have a real part which will occur if the integral in (10) has an imaginary part. The integral in (10) does have imaginary parts coming from the contour integrations involving the poles in the integrand at $s_n = n\pi/eE_{EM}$. (The $1/s^2$ divergence in the integrand in (10) leads to an infinite imaginary part which is removede via renormalization [7]). Taking semi-circular deviations around each of the poles s_n leads to a real part of ζ given by

$$\begin{aligned}
\gamma = Re(\zeta) \quad &= \quad L^3 T \frac{e^2 E_{EM}^2}{16\pi^3} \sum_{n=1}^{\infty} \frac{(-1)^{n+1}}{n^2} \exp\left(-\frac{\pi n m^2}{eE_{EM}}\right) \\
&\approx \quad L^3 T \frac{e^2 E_{EM}^2}{16\pi^3} \exp\left(-\frac{\pi m^2}{eE_{EM}}\right) \ .
\end{aligned} \tag{11}$$

Due to the exponential term the main contribution comes from $n = 1$ term in the sum. The result in (11) can be used to obtain the result in (1) by inserting it into (7) and dividing by (Vol × time)

$$\begin{aligned}
\frac{\text{prob}_{e^+e^-}}{\text{Vol} \times \text{time}} &= \frac{1}{\text{Vol} \times \text{time}} (1 - e^{-2\gamma}) \approx \frac{e^2 E_{EM}^2}{8\pi^3 \hbar c} \exp\left(-\frac{\pi m^2 c^3}{eE_{EM} \hbar}\right) \\
&= \frac{\alpha_{EM} E_{EM}^2}{8\pi^3 \hbar} \exp\left(-\frac{\pi m^2 c^3}{eE_{EM} \hbar}\right) \ ,
\end{aligned} \tag{12}$$

[2] It is only the discrete part of λ_n (*i.e.* $eE_0(2n+1)$ which is summed over. The "sum" over the momenta part of λ_n (*i.e.* p_x, p_y and p_z) involve integrations. The details leading to the result in (10) are in reference [3], and in addition we perform essentially the same steps in appendix A when we calculate the production of gluons in a uniform chromoelectric field.

where we have restored factors of $\hbar$ and c and then written the result in terms of the fine structure constant $\alpha_{EM} = \frac{e^2}{\hbar c} \approx \frac{1}{137}$. We want to use (12) to determine $\text{prob}_{e^+e^-}$ when the electric field is at the critical value $E_{EM} \approx 1.4 \times 10^{14} \frac{dyne}{esu}$. To do this we need some way to characterize what a "natural" value is for the volume and time in the denominator of (12). For the volume we take it to be the cube of the reduced Compton wave length of the electron $\text{Vol} = \left(\frac{\hbar}{mc}\right)^3 \approx 5.7 \times 10^{-32} cm^3$. For the time we take the reduced Compton time $\text{time} = \frac{\hbar}{mc^2} \approx 1.3 \times 10^{-21} sec$. Putting all these values together in (12) yields

$$\text{prob}_{e^+e^-} \approx 0.015 \tag{13}$$

In the next section we will calculate the equivalent result to equation (12) but for gluons produced from uniform chromoelectric fields and uniform chromomagnetic fields. We will set the probability of gluon production (*i.e.* prob_{gg}) to the same value as that of e^+e^- production from (13) for the critical electric field, E_{EM}. In this way we will arrive at the critical chromoelectric/chromomagnetic field strength to produce gluons. The initial expectation might be that the critical color field strength might be smaller since gluons are massless so there will be no exponential suppression as occurs due to the rest mass of the electron for the QED case.

3. Schwinger effect for uniform color electric field and uniform color magnetic field

To calculate the production rate for gluons from uniform chromoelectric and chromomagnetic fields we follow reference [8] and place gluon field excitations in a fixed background field. For the two background fields we take a uniform chromoelectric field and uniform chromomagnetic field. These calculations are rather involved and we put the details in appendix A for the uniform chromoelectric field and appendix B for the uniform chromomagnetic field. The result for the production rate for gluons in a uniform chromoelectric field is given by equation (A.26) as

$$\gamma \approx 0.00187 \times L^3 T \frac{g^2 E_0^2}{16\pi^3} \, . \tag{14}$$

Using this result from (14) in (7) we obtain the probability for SU(2) gluon creation per unit volume and unit time

$$\frac{\text{prob}_{gg}}{\text{Vol} \times \text{time}} = \frac{1}{\text{Vol} \times \text{time}} (1 - e^{-2\gamma}) \approx \frac{0.00187 g^2 E_0^2}{8\pi^3 \hbar c} = \frac{0.00187 \alpha_{QCD} E_0^2}{8\pi^3 \hbar} \, . \tag{15}$$

In the last step we have restored factors of $\hbar$ and c and introduced the QCD fine structure constant $\alpha_{QCD} = \frac{g^2}{\hbar c}$. We will take $\alpha_{QCD} \approx 1$ so that we are in non-perturbative regime. For the present case we take the distance scale to be $10^{-12} cm$. This is one order of magnitude larger than the typical strong interaction bound state size of 1fermi. Thus is the volume factor in (15) we have $\text{Vol} = 10^{-36} cm^3$ and for the time factor we take $\text{time} = \frac{10^{-12}}{3 \times 10^{10}} = 3.3 \times 10^{-23}$sec. Finally we take the value for the probability of gluon production, prob_{gg} at the critical chromoelectric field magnitude to be the same as that for the electromagnetic case as given in (13) namely $\text{prob}_{\text{gg}} \approx 0.015$, using these assumptions in (15) we can calculate the value of the critical chromoelectric field magnitude as

$$E_0 \approx 2.5 \times 10^{17} \frac{\text{dyne}}{\text{esu}} \, . \tag{16}$$

This is then three orders of magnitude larger than the $E\&M$ critical field value of $E_{EM} \approx 1.4 \times 10^{14} \frac{dyne}{esu}$. Thus although gluons are massless (and thus there is no exponential suppression factor as in the electromagnetic case due to the mass of the electron) this does not lower the critical chromoelectric field value, and in fact we find the critical field value for QCD is three orders of magnitude larger than in the $E\&M$ case. This, at first sight, surprising result arises from the fact that QCD has a confinement scale in the range of 1fermi $= 10^{-13}$cm. This distance is smaller than the Compton wave length of the electron which we used to set the volume and time in the electromagnetic case as Vol $= \left(\frac{\hbar}{mc}\right)^3 \approx 5.7 \times 10^{-32} cm^3$ and time $= \frac{\hbar}{mc^2} \approx 1.3 \times 10^{-21} sec$ respectively. In the QCD case we overestimated the distance scale as 10 times larger than 1fermi $= 10^{-13}$cm and also used this to obtain the characteristic time for the QCD case. Even though we (slightly) overestimated the QCD distance and time scales (which would by (15) would be favorable toward to decreasing the critical QCD field strength) we nevertheless found that the critical QCD field strength was larger than the critical $E\&M$ critical field strength. Even if we had used (without justification) the electron Compton wave length and associated Compton time in (15) this would give a QCD critical field strength of $E_0 \sim 10^{14} \frac{dyne}{esu}$ *i.e.* the same as for $E\&M$. Thus the masslessness of the gluons does not lead to a lower critical field strength relative to what is found in the canonical Schwinger mechanism case.

4. Conclusion

We presented the Schwinger effect for SU(2) gluons. We did this since gluons being massless we expect this effect to be more important than standard Schwinger effect which is exponentially suppressed due to electron/positron rest mass. The calculation details for gluon production from a uniform chromoelectric field and from a uniform chromomagnetic field are found in appendix A and B respectively. As in the electromagnetic case a chromoelectric field will produce gluons while the chromomagnetic field does not.

The motivation for investigating gluon production, as opposed to the production of quarks, was that since gluons are massless there will not be an exponential suppression due to the mass of the quarks that one finds in the electromagnetic case due to the mass of the electron (*i.e.* the last exponential term in (1). However, due to confinement the natural length scales and time scale for the QCD case were several orders of magnitude smaller than the similar scales for the electromagnetic case which were set to the reduced Compton length of the electron and the associated time. This in turn made the factor Vol $\times$ time smaller for the QCD case as compared to the $E\&M$ case which in turn, given the expression for the probability for production per unit volume per unit time in (15) gave a larger magnitude for the chromoelectric field for producing gluons as compared to the magnitude of the electric field for producing electrons and positrons.

Acknowledgments

DS is supported by grant Φ.0755 in fundamental research in Natural Sciences by the Ministry of Education and Science of Kazakhstan.

Appendix A. Uniform Color Electric Field Calculation

To have a constant color electric field in the $\hat{z} = 3$ spatial direction and in the $a = 3$ color direction,

$$E^a_\mu \rightarrow E^3_3 = F^3_{03} = \partial_0 A^3_3 - \partial_3 A^3_0 + g\epsilon^{3bc} A^b_3 A^c_0 = E_0 \hat{z} \ , \tag{A.1}$$

there are two gauge choices one can make for the potential

$$(i) \ \ A^a_\mu = E_0 t \delta_{\mu 3} \delta^{a3} \quad ; \quad (ii) \ \ A^a_\mu = -E_0 z \delta_{\mu 0} \delta^{a3} \tag{A.2}$$

For the calculation in this appendix we use the (i) form of the potential. Inserting form (i) from (A.2) into (A.1) does give $E^a_\mu \rightarrow E^3_3 = \partial_0 A^3_3 = E_0 \hat{z}$. We now take the potential (i) from (A.2) as a background potential (*i.e.* $A^{(0)a}_\mu$) and we consider small variation A^a_ν around this background (the background nature of the potential is indicated by the superscript (0)). In this way the QCD action can be written as

$$F^a_{\mu\nu}F^{a\mu\nu} = \left[\left(\partial_\mu A^{(0)a}_\nu - \partial_\nu A^{(0)a}_\mu\right) + \left(\partial_\mu A^a_\nu - \partial_\nu A^a_\mu\right) + g\epsilon^{abc}\left(A^{(0)b}_\mu + A^b_\mu\right)\left(A^{(0)c}_\nu + A^c_\nu\right)\right]^2 \quad \text{(A.3)}$$

We now run through the color indices $a = 1, 2, 3$ and insert the explcit form of the background potential $A^{(0)3}_\mu = E_0 t\delta_{\mu 3}$. The only non-zero values of $\partial_\mu A^{(0)3}_\nu - \partial_\nu A^{(0)3}_\mu + g\epsilon^{3bc}A^{(0)b}_\mu A^{(0)c}_\nu$ are when $\mu = 0$ and $\nu = 3$ or vice versa, each of which contribute a term of E_0.

$$\begin{aligned} F^a_{\mu\nu}F^{a\mu\nu} &= \left\{\partial_\mu A^1_\nu - \partial_\nu A^1_\mu + g\left[A^2_\mu\left(A^3_\nu + E_0 t\delta_{\nu 3}\right) - \left(A^3_\mu + E_0 t\delta_{\mu 3}\right)A^2_\nu\right]\right\}^2 \\ &+ \left\{\partial_\mu A^2_\nu - \partial_\nu A^2_\mu + g\left[\left(A^3_\mu + E_0 t\delta_{\mu 3}\right)A^1_\nu - A^1_\mu\left(A^3_\nu + E_0 t\delta_{\nu 3}\right)\right]\right\}^2 \\ &+ \left[\partial_\mu A^3_\nu - \partial_\nu A^3_\mu + g\left(A^1_\mu A^2_\nu - A^2_\mu A^1_\nu\right)\right]^2 \\ &+ 2E_0^2 - 2E_0\left[\partial_\mu A^3_\nu - \partial_\nu A^3_\mu + g\left(A^1_\mu A^2_\nu - A^2_\mu A^1_\nu\right)\right]\left(\delta_{\mu 3}\delta_{\nu 0} - \delta_{\mu 0}\delta_{\nu 3}\right) \end{aligned} \quad \text{(A.4)}$$

We now expand the above expression to 2nd order in the A^a_μ which gives

$$\begin{aligned} F^a_{\mu\nu}F^{a\mu\nu} &\approx \left[\partial_\mu A^1_\nu - \partial_\nu A^1_\mu - gE_0 t\left(A^2_\nu\delta_{\mu 3} - A^2_\mu\delta_{\nu 3}\right)\right]^2 \\ &+ \left[\partial_\mu A^2_\nu - \partial_\nu A^2_\mu - gE_0 t\left(A^1_\mu\delta_{\nu 3} - A^1_\nu\delta_{\mu 3}\right)\right]^2 \\ &+ \left(\partial_\mu A^3_\nu - \partial_\nu A^3_\mu\right)^2 + 2E_0^2 - 4E_0\left[\partial_3 A^3_0 - \partial_0 A^3_3 + g\left(A^1_3 A^2_0 - A^2_3 A^1_0\right)\right] \end{aligned} \quad \text{(A.5)}$$

We now require that at spatial and temporal infinity the variation of the potential goes to zero $A^3_0\left(x_3 \to \pm\infty\right) = 0$ and $A^3_3\left(x_0 \to \pm\infty\right) = 0$. This causes the source terms for A^3_μ in the last term to vanish by partial integration. Note that $\left(\partial_\mu A^3_\nu - \partial_\nu A^3_\mu\right)^2$ is a free Lagrange density, so it is not of interest and can be dropped.

$$\begin{aligned} F^a_{\mu\nu}F^{a\mu\nu} &= \left[\partial_\mu A^1_\nu - \partial_\nu A^1_\mu - gE_0 t\left(A^2_\nu\delta_{\mu 3} - A^2_\mu\delta_{\nu 3}\right)\right]^2 \\ &+ \left[\partial_\mu A^2_\nu - \partial_\nu A^2_\mu - gE_0 t\left(A^1_\mu\delta_{\nu 3} - A^1_\nu\delta_{\mu 3}\right)\right]^2 \\ &+ 2E_0^2 - 4E_0 g\left(A^1_3 A^2_0 - A^2_3 A^1_0\right) \end{aligned} \quad \text{(A.6)}$$

We now re-write this using complex potential forms as

$$\begin{aligned} F^a_{\mu\nu}F^{a\mu\nu} &= \left|\partial_\mu\left(A^1_\nu + iA^2_\nu\right) - \partial_\nu\left(A^1_\mu + iA^2_\mu\right) - gE_0 t\left[\left(A^2_\nu - iA^1_\nu\right)\delta_{\mu 3} - \left(A^2_\mu - iA^1_\mu\right)\delta_{\nu 3}\right]\right|^2 \\ &+ 2E_0^2 - i2gE_0\left[\left(A^1_3 + iA^2_3\right)\left(A^1_0 - iA^2_0\right) - \left(A^1_0 + iA^2_0\right)\left(A^1_3 - iA^2_3\right)\right] \end{aligned} \quad \text{(A.7)}$$

We now replace A^1_μ and A^2_μ by defining the following complex potetnials

$$W_\mu = \frac{1}{\sqrt{2}}\left(A^1_\mu + iA^2_\mu\right) \quad , \quad W^\dagger_\mu = \frac{1}{\sqrt{2}}\left(A^1_\mu - iA^2_\mu\right). \quad \text{(A.8)}$$

In terms of these new, complex potentials $F^a_{\mu\nu}F^{a\mu\nu}$ becomes

$$\begin{aligned} F^a_{\mu\nu}F^{a\mu\nu} &= \left|\sqrt{2}\partial_\mu W_\nu - \sqrt{2}\partial_\nu W_\mu - i\sqrt{2}gE_0 tW_\nu\delta_{\mu 3} + i\sqrt{2}gE_0 tW_\mu\delta_{\nu 3}\right|^2 \\ &+ 2E_0^2 - i2gE_0\left(2W^\dagger_3 W_0 - 2W^\dagger_0 W_3\right) \end{aligned} \quad \text{(A.9)}$$

IOP Conf. Series: Journal of Physics: Conf. Series **883** (2017) 012014 doi:10.1088/1742-6596/883/1/012014

This finally leads to the following Lagrange density for gluons in the background potential of a uniform color electric field

$$\begin{aligned}\mathcal{L} &= -\frac{1}{4}F^a_{\mu\nu}F^{a\mu\nu} \\ &= -\frac{1}{2}\left|(\partial_\mu - igE_0 t\delta_{\mu 3})W_\nu - (\partial_\nu - igE_0 t\delta_{\nu 3})W_\mu\right|^2 - igE_0\left(W_3^\dagger W_0 - 2W_0^\dagger W_3\right) - \frac{1}{2}E_0^2\end{aligned} \tag{A.10}$$

This Lagrange density leads to the following equation of motion

$$\left(\partial^\mu - igE_0 t\delta^{\mu 3}\right)\left[(\partial_\mu - igE_0 t\delta_{\mu 3})W_\nu - (\partial_\nu - igE_0 t\delta_{\nu 3})W_\mu\right] - igE_0\left(\delta_{\nu 3}W_0 - \delta_{\nu 0}W_3\right) = 0. \tag{A.11}$$

Choosing the background gauge condition $(\partial^\mu - igE_0 t\delta^{\mu 3})W_\mu = 0$ [3] simplifies (A.11) to

$$\left(\partial^\mu - igE_0 t\delta^{\mu 3}\right)(\partial_\mu - igE_0 t\delta_{\mu 3})W_\nu = -2igE_0\left(W_0\delta_{\nu 3} - W_3\delta_{\nu 0}\right) \tag{A.12}$$

The above equation can be written in matrix form as

$$\left(\partial^\mu - igE_0 t\delta^{\mu 3}\right)^2 W_\nu - 2gE_0\begin{bmatrix} 0 & 0 & 0 & -i \\ 0 & 0 & 0 & 0 \\ 0 & 0 & 0 & 0 \\ i & 0 & 0 & 0 \end{bmatrix} W_\mu = 0 \tag{A.13}$$

The four eigenvalues for this matrix are ± 1 and a double eigenvalue of zero. The zero eigenvalues are excluded by our chosen gauge condition so for now we consider just the eigenvalues ± 1. Thus (A.13) becomes

$$\left[\left(\partial^\mu - igE_0 t\delta^{\mu 3}\right)^2 W_\nu - 2gE_0(\pm 1)\right] W_\nu = 0. \tag{A.14}$$

Recalling that we are using the metric signature $(+,-,-,-)$ (A.14) can be expanded as

$$\left\{\frac{\partial^2}{\partial t^2} - \frac{\partial^2}{\partial x_1^2} - \frac{\partial^2}{\partial x_2^2} + \left(i\frac{\partial}{\partial x_3} + gE_0 t\right)^2 \mp 2gE_0\right\} W_\nu = 0 \tag{A.15}$$

Next we Fourier transform (A.15) to $\tilde{W}_\nu(t, k_1, k_2, k_3)$ with the result

$$\left\{\frac{\partial^2}{\partial t^2} + k_1^2 + k_2^2 + g^2E_0^2\left(\frac{k_3}{gE_0} + t\right)^2 \pm 2gE_0\right\}\tilde{W}_\nu = 0 \tag{A.16}$$

Now making the substitution $t' = \dfrac{k_3}{gE_0} + t$ (for which one has $\frac{\partial}{\partial t} = \frac{\partial}{\partial t'}$) and then performing a rotaion to imaginary time, $t' \to -i\tau$, $E_0 \to -iE_0$ we arrive at

$$\left\{-\frac{\partial^2}{\partial \tau^2} + g^2E_0^2\tau'^2 + k_1^2 + k_2^2 \pm 2igE_0\right\}\tilde{W}_\nu = 0 \tag{A.17}$$

[3] This background gauge condition singles out the physical degrees of freedom. Formally one can show this by using the gauge condition to find the corresponding ghost fields and showing that these cancel the contribution of A^3_μ and those components of W_μ for which $\left(\partial^\mu - igE_0 t\delta^{\mu 3}\right)W_\mu \neq 0$. Thus although ghost fields do not appear explicitly in our calculations they are taken into account implicitly via the background gauge condition and our taking $A^3_0(x_3 \to \pm\infty) = 0$ and $A^3_3(x_0 \to \pm\infty) = 0$.

The first two terms in (A.17) correspond to a harmonic oscillator with frequency $\omega = gE_0$, which has eigenvalues given by $-\frac{\partial}{\partial\tau^2} + g^2E_0^2\tau^2 \rightarrow 2\left(n+\frac{1}{2}\right)gE_0$. With this (A.17) becomes $\{(2n+1)\,gE_0 + k_1^2 + k_2^2 \pm igE_0\}\,\tilde{W}_\nu = 0$. From this equation we can read of the eigenvalues as

$$\lambda_n = (2n+1)\,gE_0 + k_1^2 + k_2^2 \pm 2igE_0. \tag{A.18}$$

Note that λ_n is a combination of discrete (*i.e.* $(2n+1)\,gE_0$) and continuous (*i.e.* $k_1^2+k_2^2$) parts. Substituting these eigenvalues from (A.18) into (5) we find that ζ becomes

$$\zeta = -\sum_n \int_0^\infty \frac{\mathrm{d}s}{s} \exp\left\{-\left[(2n+1)\,gE_0 + k_1^2 + k_2^2 \pm 2igE_0\right]s\right\} \tag{A.19}$$

We take our system to be quantized in a cubical spatial volume with sides of length L and over a total (imaginary) interaction time $\tau = -iT$ which then turns (A.19) into

$$\zeta = -L\int_{-\infty}^{\infty}\frac{\mathrm{d}k_1}{(2\pi)}L\int_{-\infty}^{\infty}\frac{\mathrm{d}k_2}{(2\pi)}L\int_0^{gE_0\tau}\frac{\mathrm{d}k_3}{(2\pi)}\sum_{n=0}^{\infty}\int_0^\infty\frac{\mathrm{d}s}{s}\exp\left\{-\left[(2n+1)\,gE_0 + k_1^2 + k_2^2 \pm 2igE_0\right]s\right\}. \tag{A.20}$$

The k_1 and k_2 integrations related to the momentum in the free directions and are simple Gaussian integrals, which give two factors of $\sqrt{\frac{\pi}{s}}$. The integration in the k_3 direction is related to the momentum in the z direction which is the direction of the chromoelectric field. Thus as in [3] this integration is constrained to the range $0 < k_3 < gE_0\tau$. Performing the k_1, k_2, k_3 integrations gives

$$\zeta = -L^3\tau\frac{gE_0}{8\pi^2}\sum_{n=0}^{\infty}\int_0^\infty\frac{\mathrm{d}s}{s^2}\exp\left\{-\left[(2n+1)\,gE_0 \pm 2igE_0\right]s\right\} \tag{A.21}$$

We note that $\sum\limits_{n=0}^{\infty} e^{-(2n+1)gE_0 s} = \frac{1}{2}\frac{1}{\sinh(gE_0 s)}$ which then transforms (A.21) into

$$\zeta = -L^3\tau\frac{gE_0}{16\pi^2}\int_0^\infty\frac{\mathrm{d}s}{s^2}\frac{e^{\pm 2igE_0 s}}{\sinh(gE_0 s)} \tag{A.22}$$

We now return to real time via the rotation $\tau \rightarrow iT$ which also involves changing the magnitude of the chromoelectric field as $E_0 \rightarrow iE_0$. This gives

$$\zeta = -iL^3T\frac{gE_0}{16\pi^2}\int_0^\infty\frac{\mathrm{d}s}{s^2}\frac{e^{\mp 2gE_0 s}}{\sin(gE_0 s)} \tag{A.23}$$

Equation (A.23) is now at the point in the electromagnetic calculation given by equation (10), but now the exponential factor involving the electron mass ($\exp[-m^2 s]$) is replaced by an exponential suppression involving the field strength ($e^{\mp 2gE_0 s}$). As before if ζ has a real part (*i.e.* if the integral in (A.23) has an imaginary contribution) there will be particle production. As in the case of the electromagnetic integral in (10) the integral in (A.23) does have an imaginary contribution coming from the poles of $\frac{1}{\sin(gE_0 s)}$ whihc are located at $s_n = \frac{n\pi}{gE_0}$, where n is an integer. As before we ignore the singularity at $s = 0$. The integration contours are infinitesimal semicircular in the upper half plane and from (A.23) this gives

$$\begin{aligned} \gamma &= \mathrm{Re}(\zeta) = -iL^3T\frac{gE_0}{16\pi^2}\sum_{n=1}^{\infty}\int_{s_n-\epsilon}^{s_n+\epsilon}\frac{\mathrm{d}s}{s^2}\frac{e^{\mp 2gE_0 s}}{\sin(gE_0 s)} \\ &= -iL^3T\frac{gE_0}{16\pi^2}\sum_{n=1}^{\infty}\int_{s_n-\epsilon}^{s_n+\epsilon}\frac{\mathrm{d}s}{s^2}\frac{e^{\mp 2gE_0 s}}{\cos(gE_0 s_n)gE_0\,(s-s_n)} \ . \end{aligned} \tag{A.24}$$

In the last step we have expanded $\sin(gE_0 s)$ around the poles at s_n. From the Residue theorem, the poles of the integral in (A.24) give $-i\pi \times \sum \mathrm{Res}(\text{function})$, where the sum is over the residue of the integrand. The result is

$$\gamma = \mathrm{Re}(\zeta) = -L^3T\frac{1}{16\pi}\sum_{n=1}^{\infty}\frac{e^{\mp 2gE_0 s_n}}{s_n^2\cos(gE_0 s_n)} = L^3T\frac{g^2E_0^2}{16\pi^3}\sum_{n=1}^{\infty}(-1)^{n+1}\frac{e^{\mp 2n\pi}}{n^2} \ , \tag{A.25}$$

where in the last step we have substitute in the poles $s_n = \frac{n\pi}{gE_0}$ and simplify. The $(-1)^{n+1}$ comes from $-\cos(n\pi)$. The $e^{+n\pi}$ choice in (A.25) leads to a divergent γ so we take the $e^{-n\pi}$ choice in (A.25) which leads to

$$\gamma = L^3T\frac{g^2E_0^2}{16\pi^3}\sum_{n=1}^{\infty}(-1)^{n+1}\frac{e^{-2n\pi}}{n^2} \approx 0.00187 \times L^3T\frac{g^2E_0^2}{16\pi^3} \ . \tag{A.26}$$

In the last step we have carried out the sum numerically with the result $\sum_{n=1}^{\infty}(-1)^{n+1}\frac{e^{-2n\pi}}{n^2} \approx 0.00187$.

Calculations similar to the above were carried out in [9, 10, 11] for gluon production in a constant $SU(3)$ color electric field. In these works the integration over the transverse momentum was not carried out, but if one does carry out the integration of the transverse momentum of the results in [9, 10, 11] one finds that our result for γ in (A.26) is consistent with these previous results. In particular there is no mass suppression (as expected) and the production rate is proportional to $g^2E_0^2$

Appendix B. Uniform Color Magnetic Field Calculation

In this appendix we show that, unlike the uniform color electric field of appendix A, a uniform color magnetic field does not produce gluons. This is similar to what occurs in the electromagnetic Schwinger effect – a uniform magnetic field does not produce electrons/positrons.

A constant color magnetic field in the $\hat{z} = 3$ spatial direction and in the $a = 3$ color direction,

$$B_\mu^a \to B_3^3 = F_{12}^3 = \partial_1 A_2^3 - \partial_2 A_1^3 + g\epsilon^{3bc}A_1^b A_2^c = H_0\hat{z} \ , \tag{B.1}$$

can be obtained by the potential

$$A_\mu^a = -H_0 x_1 \delta_{\mu 2}\delta^{a3} \ . \tag{B.2}$$

With the potential in (B.2) we perform a long calculation that is similar to the one for the constant color electric field, which leads to

$$\left(\partial^\mu - igH_0 x\delta^{\mu 2}\right)^2 W_\nu - 2gH_0\begin{bmatrix} 0 & 0 & 0 & 0 \\ 0 & 0 & i & 0 \\ 0 & -i & 0 & 0 \\ 0 & 0 & 0 & 0 \end{bmatrix} W_\mu = 0 \ . \tag{B.3}$$

which is the color magnetic version of (A.13). Expanding (B.3) out we arrive at the color magnetic version (A.15) which leads to

$$\left[\frac{\partial^2}{\partial t^2} - \frac{\partial^2}{\partial x^2} - \frac{\partial^2}{\partial z^2} \mp 2gH_0 - \left(i\frac{\partial}{\partial y} + gH_0 x\right)^2\right] W_\nu = 0 \tag{B.4}$$

Next we Fourier transform (B.4) to $\tilde{W}_\nu(t, k_1, k_2, k_3)$ gives the color magnetic version of (A.16)

$$\left[-E^2 + k_3^2 \mp 2gH_0 - \frac{\partial^2}{\partial x^2} + g^2 H_0^2 \left(x + \frac{k_2}{gH_0}\right)^2\right] \tilde{W}_\nu = 0\ . \tag{B.5}$$

Making the change of variable $\eta = x + \frac{k_2}{gH_0}$ this becomes

$$\left[-E^2 + k_3^2 \mp 2gH_0 - \frac{\partial^2}{\partial \eta^2} + g^2 H_0^2 \eta^2\right] \tilde{W}_\nu = 0\ . \tag{B.6}$$

The $-\frac{\partial^2}{\partial \eta^2} + g^2 H^2 \eta^2$ part of the above equation is the equation for a simple harmonic oscillator which has eignevalues $2\left(n + \frac{1}{2}\right) gH_0$. Substituting this eigenvalue into (B.6) leads to

$$\left[-E^2 + k_3^2 \mp 2gH_0 + 2\left(n + \frac{1}{2}\right) gH_0\right] \tilde{W}_\nu = 0 \tag{B.7}$$

We now rotate to imaginary time (Euclidean) as $t \to -i\tau$ and in conjunction with this we rotate the energy as $E \to= -iE$. This turns (B.7) into

$$\left[E^2 + k_3^2 \mp 2gH_0 + 2\left(n + \frac{1}{2}\right) gH_0\right] \tilde{W}_\nu = 0 \tag{B.8}$$

The eigenvalues for the system can then be written out as

$$\lambda_n = E^2 + k_3^2 \mp 2gH_0 + 2\left(n + \frac{1}{2}\right) gH_0 \tag{B.9}$$

Now substituting (B.9) into (5) yields ζ for this case as

$$\zeta = -\sum_n \int_0^\infty \frac{\mathrm{d}s}{s} \exp\left\{-\left[E^2 + k_3^2 + (2n+1)gH_0 \mp 2gH_0\right] s\right\} \tag{B.10}$$

Recalling that we are quantizing in a cube with sides length L and over a total (imaginary) interaction time $\tau = -iT$ transforms (B.10) into [4]

$$\zeta = -\tau \int_{-\infty}^{\infty} \frac{\mathrm{d}E}{(2\pi)} L \int_{-\infty}^{\infty} \frac{\mathrm{d}k_3}{(2\pi)} L^2 \frac{gH_0}{2\pi} \sum_{n=0}^{\infty} \int_0^{\infty} \frac{\mathrm{d}s}{s} \exp\left\{-\left[E^2 + k_3^2 + (2n+1)gH_0 \mp 2gH_0\right] s\right\}\ . \tag{B.11}$$

[4] From equation 8.25 in [8] the integral over the momenta in the x and y directions is $\int \frac{\mathrm{d}k_1 \mathrm{d}k_2}{(2\pi)^2} \to \frac{\int k\mathrm{d}k}{2\pi} \to \frac{gH_0}{2\pi} \sum_n$

The E and k_3 integrations are simple Gaussian integrals that give two factors of $\sqrt{\frac{\pi}{s}}$. Thus (B.11) simplifies to

$$\zeta = -\tau L^3 \frac{gH_0}{8\pi^2} \sum_{n=0}^{\infty} \int_0^{\infty} \frac{\mathrm{d}s}{s^2} \exp\left\{-\left[(2n+1)gH_0 \mp 2gH_0\right] s\right\} \ . \tag{B.12}$$

We now use the geometric series $\sum_{n=0}^{\infty} e^{-(2n+1)gHs} = \frac{1}{2}\frac{1}{\sinh(gHs)}$ to write (B.12) as

$$\zeta = -\tau L^3 \frac{gH_0}{16\pi^2} \int_0^{\infty} \frac{\mathrm{d}s}{s^2} \frac{e^{\pm 2gH_0 s}}{\sinh(gH_0 s)} \tag{B.13}$$

We now return to real time via the rotation $\tau \to iT$.

$$\zeta = -iTL^3 \frac{gH_0}{16\pi^2} \int_0^{\infty} \frac{\mathrm{d}s}{s^2} \frac{e^{\pm 2gH_0 s}}{\sinh(gH_0 s)} \tag{B.14}$$

In contrast to the color electric case of appendix A where the rotation back to real time, $\tau \to iT$, was accompanied by a rotation of the electric field amplitude, $E_0 \to iE_0$, here the rotation back to real time does not lead to a change in the color magnetic field amplitude. In comparing (B.14) with (A.23) one finds that (B.14) does not have the series of poles at $s_n = n\pi/qE_0$ that are found in (A.23). Therefore the integral in (B.14) does not have any imaginary contributions coming from the residue theorem was was the case for the integral in (A.23). Thus from (B.14) the real part of ζ is zero so $Re(\zeta) = \gamma = 0$. Thus in the case of a uniform color magnetic field there is no particle production as is expected form the electromagnetic case where one finds that a constant magnetic field does not produce electron/positron pairs.

References

[1] J. Schwinger, Phys. Rev. **82**, 664-679 (1951) .
[2] W. Heisenberg and H. Euler, Z. Physik **98**, 714 (1936).
[3] B. R. Holstein, Am. J. Phys. **67**, 499-507 (1999).
[4] K. Srinivasan and T. Padmanabhan, Phys. Rev. **D60**, 024007 (1999).
[5] R. Brout, R. Parentani, and Ph. Spindel, Nucl. Phys. **B353**, 209-236 (1991).
[6] F. Gelis and N. Tanji, Prog. Part. Nucl. Phys. **87**, 1-49 (2016).
[7] B.R. Holstein, *Topics in Advanced Quantum Mechanics* (Addison-Wesley, Reading, MA, 1992) Chap. III.6
[8] W. Greiner and A. Schäfer, *Quantum Chromodynamics* (Springer-Verlag, Berlin 1994) Chap. 8.1
[9] G. C. Nayak, P. van Nieuwenhuizen, Phys.Rev. **D71**, 125001 (2005).
[10] F. Cooper and G. C. Nayak, Phys.Rev. **D73**, 065005 (2006).
[11] G. C. Nayak, Eur. Phys. J. **C59**, 715 (2009).

Gauging away a big bang

Chethan Krishnan and Avinash Raju

Centre for High Energy Physics, Indian Institute of Science, Bangalore, India. 560012

E-mail: chethan.krishnan@gmail.com,avinashraju777@gmail.com

Abstract. We argue that in the tensionless phase of string theory where the stringy gauge symmetries are unbroken, (at least some) cosmological singularities can be understood as gauge artefacts. We present two conceptually related, but distinct, pieces of evidence: one relying on spacetime and the other on worldsheet.

1. Cosmological Singularities in String Theory

Let us start where we often start: namely, with the observation that an infinite number of independent UV divergences arise when one tries to quantize gravity perturbatively. This infinite number of undetermined couplings means that we need to do an infinite number of experiments to determine them, and this results in a loss of predictivity. This is one reason why we believe that General Relativity must require modifications at short distances.

String theory claims to solve this problem. The way it solves it, is via an enormous gauge symmetry on the worldsheet, called conformal invariance. Even though superficially the theory contains an infinite number of new particles and their couplings, making the situation *prima facie* worse, this gauge symmetry is so big that it relates *all* of the new couplings to each other. Indeed, one is left with no independent parameters at all in string theory. This makes the theory predictive (in principle) at all scales[1]. This is the jumbo version of a familiar phenomenon from particle physics: the UV breakdown of Fermi theory at the Weak scale requires the introduction of new degrees of freedom, but the new couplings are fixed by the (Higgsed) $SU(2) \times U(1)$ gauge symmetry ensuring that the theory does not give rise to an uncontrollable number of new divergences.

The worldsheet gauge symmetry of string theory will be an important ingredient in our discussions in this article.

The perturbative non-renormalizability discussed above is a *quantum* reason why we believe gravity needs modifications at short distances. But there is also a well-known *classical* reason. The singularity theorems of Penrose and Hawking [1] necessitate that in Einstein gravity, regular Cauchy data can evolve and become singular (in the past or future). This is problematic, if we want our theory to be a predictive model of Nature even in the most rudimentary sense.

One might take consolation in the fact that black hole singularities are covered up by event horizons in classical gravity, at least if Cosmic Censorship holds true[2]. But even this grim

[1] At least after one figures out which solution of the theory corresponds to our Universe! These solutions however can unfortunately be too numerous – this is the so-called String Landscape with its associated necessity for anthropic arguments.

[2] Of course, the information paradox comes back to bite us at the quantum level, even if this claim was correct!

consolation, at best only holds for an asymptotic observer. An infalling observer still will face loss of predictivity at the future singularity inside the black hole, which he/she is guaranteed to hit. Further, fine-tuned initial data can result in nakedly singular evolution even in classical gravity. Even more relevant for us, the cosmic singularity in our past is naked, and that is something we cannot help but confront.

Since singularities are believed to be short distance features, they point towards a UV modification of gravity. String theory is expected to be UV-finite, so one might hope that it can help cure singularities. Indeed, stringy resolutions of some singularities do exist [2, 3, 4, 5, 6], but these are at the level of case studies. It is fair to say that a coherent overall picture of singularities in string theory is lacking.

In particular, cosmological singularities are doubly hard in string theory. This is because cosmology means time-dependence, and quantizing strings in time-dependent backgrounds is difficult. Typically only supersymmetric backgrounds are under control, and supersymmetry forces time *in*-dependence on us.

The question of cosmic singularities in string theory then, at least as we have phrased it above, seems impossible. But one way forward is known. This is to consider timelike/null quotients of Minkowski space as simple examples of singular cosmologies. The tractability of string theory in the covering flat space enables us to probe these singular geometries using strings via orbifold techniques. Two popular examples where this strategy has been explored are the Milne orbifold and the null orbifold [7, 8].

However, all is not well. Tree level string scattering amplitudes on these geometries [9, 10, 11, 12, 13, 14] have problematic divergences, so it is unclear what the punchline of this line of work is. The current philosophy [15] takes the amplitude at face value and interprets the divergences as due to black hole formation when the string is inside its own Schwarzschild radius: ie., there is large backreaction from the pointlike ($\alpha' \to 0$) limit of strings.

This begs the question: what can be learnt about cosmic singularities from the opposite long-and-floppy limit ($\alpha' \to \infty$) of strings? This will be the theme of this article, and we will attempt to answer this question from two complementary angles.

2. Higher Spins and Stringy Gauge Invariance

Vasiliev has constructed interacting theories involving gravity and higher spins [16]. There is evidence that these are related to the $\alpha' \to \infty$ limit of string theory [17, 18, 19, 20]. Higher spin theory (roughly speaking) captures the spacetime dynamics of the worldsheet spectrum of states of the string, in the limit where the string states are massless (masses are inversely related to α').

The presence of extra massless modes suggests that higher spin theory has a lot more gauge invariances than general relativity, which only has diffeomorphisms. Indeed, again loosely, we can think of these higher spin gauge symmetries as the target space realization of worldsheet gauge invariance: this is the un-Higgsed phase of string theory.

We know that diffeomorphisms (freedom to change coordinates) cannot remove singularities, but these bigger gauge invariances might change that. Could some of the spacetime singularities be just artifacts of a choice of gauge in string theory? Can a higher spin gauge transformation put Milne/Null orbifold in a non-singular gauge? For concreteness, we will consider Milne in this article [21].

3. Higher Spin Resolution of a Toy Big-Bang

We will work with 3D flat space higher spin theory [22, 23]. AdS gravity in 3D has a tractable Chern-Simons formulation and this can be generalized [24] to higher spins by changing the gauge group from $SL(2)\times SL(2)$ to $SL(N)\times SL(N)$. By allowing the gauge fields to take Grassmann

IOP Conf. Series: Journal of Physics: Conf. Series **883** (2017) 012015 doi:10.1088/1742-6596/883/1/012015

values [25] one has a convenient trick[3] to handle flat space higher spins as well. This enables us to transport much of the AdS technology with little effort to the flat space case. Because the orbifolds we will consider are quotients of 2D flat space, we can always embed them in 3D.

We take the Milne metric as [27]:

$$ds^2 = -dT^2 + r_C^2 dX^2 + \alpha^2 T^2 d\phi^2, \tag{1}$$

X is noncompact and $\phi \sim \phi + 2\pi$. Spacetime is a double cone with a causal singularity at $T = 0$ where ϕ-circle crunches to a point before big-bang-ing. This metric is a solution of Einstein equations – trivially, because it is locally Ricci-flat. We can translate (1) into the $SL(2)$ Chern-Simons language where it corresponds to a flat connection that is invariantly characterized by its holonomy.

But we can think of Milne also as a solution of the higher spin, ie. $SL(N)$, version of Chern-Simons theory. Now we ask: can one do a gauge transformation involving higher spin charges that preserves the holonomy, while changing the metric so that the singularity is removed? The answer found in [21] was that indeed this is possible. It was found that (1) is gauge-equivalent to

$$ds^2 = -dT^2 + r_C^2 dX^2 + (\alpha^2 T^2 + \beta^2) d\phi^2, \tag{2}$$

where β^2 is positive: the crunch-bang singularity is now replaced by a smooth bounce, see figure 1. In the new gauge there are non-vanishing higher spin fields in the solution which can

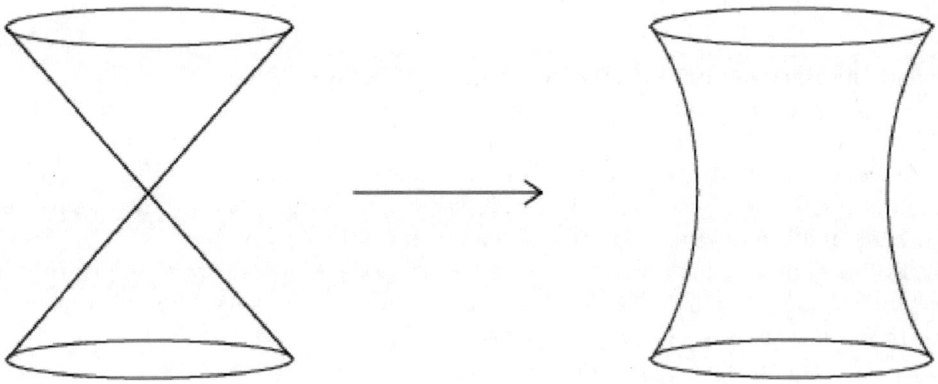

Figure 1.

be thought of as the "matter" supporting the throat[4] The curvature invariants as well as the higher spin fields remain finite everywhere. The result in [21] was found by turning on spin 3 charges, it was successfully generalized to even higher spins in [28].

Modulo some caveats which we will clarify later, this result is suggestive that the Milne singularity is a gauge artefact.

[3] We call it a trick, but the fact that many things work quite naturally and nicely under the Grassmann approach makes one wonder whether there is more to it than a trick. In particular, general Fefferman-Graham-Banados solutions [25], asymptotic symmetry algebras [25], central charges [25], generalizations to higher spins [25], addition of chemical potentials [26] all map conveniently under the Grassmann map. The immediate reason for its success is the fact that $ISO(2,1)$, the gauge group for the Chern-Simons theory that describes flat space gravity is the Inonu-Wigner contraction of $SL(2) \times SL(2)$, the corresponding gauge group in AdS_3.

[4] This might seem like a violation of singularity theorems, but it is not. Geodesics are no longer good probes in higher spin geometry; the energy conditions (for example) are gauge-*variant*. So geodesic incompleteness arguments do not apply.

IOP Conf. Series: Journal of Physics: Conf. Series **883** (2017) 012015 doi:10.1088/1742-6596/883/1/012015

4. Singularities in the Tensionless String Limit

Since higher spin theory should capture aspects of the tensionless limit of string theory, our result in the previous section suggests that the divergences in the string amplitudes that people have observed previously should not arise in this limit. The string amplitude, afterall, is a gauge-invariant object. This is what we discuss in this section.

In order to investigate the tensionless (ie. $\alpha' \to \infty$) limit of string amplitudes, we need a dimensionless α'. We are working with quotients of flat space, the only available dimensionless α' has to be constructed from the the momenta of the scattering states – there is no background scale. This is different from the AdS case, where the AdS radius works as a one-parameter family along which we can send the dimensionless α' to ∞.

The 4-pt string scattering amplitudes in Milne are complicated beasts even for the lowest lying states, but it turns out that two key results hold [29]:

- All pathological UV divergences related to the singularity arise only when $\alpha'(\mathbf{p}_1-\mathbf{p}_3)^2 \leq 2$ or similar conditions hold. Here $\mathbf{p_i}$ label the momenta of the string scattering states. Roughly speaking the α' is being measured in units of momentum transfer (in t or u channel) and when it is large enough, there are no pathological divergences.
- An exhaustive scan reveals that all other divergences are sensible IR divergences. Eg: a whole sequence of logarithmic divergences which have interpretation as the tower of intermediate string states going on-shell.

This result is re-assuring, because it is a strong suggestion that the resolvability of the Milne singularity we saw in higher spin theory is a reflection of its resolvability in the underlying string theory.

5. Towards Singularities in the Higgsed Phase of String Theory

Cosmological de Sitter quotients have been resolved previously using higher spins [30, 31]. Also, in AdS_3 it was noticed that higher spin horizons are gauge-dependent [32, 33]. But in both, making a direct connection to the string worldsheet was hard. But Minkowski orbifolds allow connections both with string theory as well as higher spins. This is the primary source of our interest in them, and the primary reason this work was possible.

We focused on Milne in this article, a parallel story holds [34] also for the null Universe of [9]. There are substantial differences in detail and technical difficulty between the two, but satisfyingly, the conclusions are essentially identical.

We found a non-singular gauge for the metric, but in these gauges we have non-trivial higher spin fields. These are finite, but typically vanish somewhere in the bulk. Is this bad? We believe this is unlikely, but to answer it unambiguously, without resorting to strings, we will need a higher spin generalization of Riemannian geometry. Since presently we do not have such a theory, we will take the well-definedness of the string amplitude at large α' as an indication that the resolution is legitimate. Also, the fact that we are able to find *some* resolution for the metric which preserves the original symmetries of the geometry, we find encouraging.

A consistent set of boundary conditions that contains both the singular (1) and the resolved (2) solutions was found recently in [26]. Happily, the gauge transformation that takes the singular geometry to our resolved geometry has zero canonical charge – ie., they are both the same state, as one would want. This is consistent: one would expect this from the string amplitude argument, because it captures gauge invariant information.

Finally, let us note that in standard general relativity one distinguishes between "coordinate" singularities and "true" singularities. The Schwarzschild horizon in the usual Schwarzschild coordinates is an example of the former: by going to a freely falling (eg: Kruskal) coordinate system at the horizon, one can see that this is an artifact of the Schwarzschild coordinates. "True" singularities on the other hand, are places beyond which geodesics cannot be extended even though their affine parameters are finite[5]. They cannot be resolved within general relativity and these were our concern in this paper. Our conclusion, at least in the un-Higgsed phase of string theory was that once the stringy gauge symmetries (and not just coordinate transformations) are taken into account, they can also be understood as artifacts of a gauge choice.

The $\alpha' \to \infty$ limit is precisely the opposite limit of the Einstein gravity limit: strings are long-and-floppy, not pointlike. What we are seeing is an explicit demonstration that the Regge limit can resolve singularities. But the message that some singularities might be gauge artefacts and might be resolved via gauge transformations is perhaps a useful paradigm to keep in mind, even away from the Regge limit. Can an approach along these lines in a Higgsed phase of string/higher-spin theory shed some light on a more realistic big-bang singularity?

Acknowledgement

This article is a summary and synthesis of various papers. We thank Ben Craps, K. Surya Kiran, Shubho Roy, Ayush Saurabh and Joan Simon for related collaborations and Arjun Bagchi, Matthias Gaberdiel, Daniel Grumiller, Shamit Kachru, Elias Kiritsis and Shiraz Minwalla for discussions, comments and/or questions that affected the final presentation. We also thank Matt Lake and Shingo Takeuchi for hospitality at Naresuan University, Phitsanulok, during the IF+YITP symposium, and APCTP for financial support.

References

[1] S. W. Hawking and G. F. R. Ellis, *The Large Sclae Structure of Spacetime*, Cambridge.
[2] L. J. Dixon, J. A. Harvey, C. Vafa and E. Witten, Nucl. Phys. B **261**, 678 (1985).
[3] L. J. Dixon, J. A. Harvey, C. Vafa and E. Witten, Nucl. Phys. B **274**, 285 (1986).
[4] A. Strominger, Nucl. Phys. B **451**, 96 (1995) [hep-th/9504090].
[5] A. Adams, J. Polchinski and E. Silverstein, JHEP **0110**, 029 (2001) [hep-th/0108075].
[6] J. McGreevy and E. Silverstein, JHEP **0508**, 090 (2005) [hep-th/0506130].
[7] G. T. Horowitz and A. R. Steif, Phys. Rev. Lett. **64**, 260 (1990).
[8] J. Simon, JHEP **0210**, 036 (2002) [hep-th/0208165].
[9] H. Liu, G. W. Moore and N. Seiberg, JHEP **0206**, 045 (2002) [hep-th/0204168].
[10] H. Liu, G. W. Moore and N. Seiberg, JHEP **0210**, 031 (2002) [hep-th/0206182].
[11] M. Berkooz, B. Craps, D. Kutasov and G. Rajesh, JHEP **0303**, 031 (2003) [hep-th/0212215].
[12] L. Cornalba and M. S. Costa, Fortsch. Phys. **52**, 145 (2004) [hep-th/0310099].
[13] B. Durin and B. Pioline, hep-th/0501145.
[14] B. Craps, Class. Quant. Grav. **23**, S849 (2006) [hep-th/0605199].
[15] G. T. Horowitz and J. Polchinski, Phys. Rev. D **66**, 103512 (2002) [hep-th/0206228].
[16] M. A. Vasiliev, Lect. Notes Phys. **892**, 227 (2015) [arXiv:1404.1948 [hep-th]].
[17] B. Sundborg, Nucl. Phys. Proc. Suppl. **102**, 113 (2001) [hep-th/0103247].
[18] E. Witten, Talk given at J.H. Schwarz 60th Birthday Conference.
[19] S. Giombi, S. Minwalla, S. Prakash, S. P. Trivedi, S. R. Wadia and X. Yin, Eur. Phys. J. C **72**, 2112 (2012) [arXiv:1110.4386 [hep-th]].
[20] M. R. Gaberdiel and R. Gopakumar, arXiv:1501.07236 [hep-th].
[21] C. Krishnan and S. Roy, Phys. Lett. B **734**, 92 (2014) [arXiv:1311.7315 [hep-th]].
[22] H. Afshar, A. Bagchi, R. Fareghbal, D. Grumiller and J. Rosseel, Phys. Rev. Lett. **111**, no. 12, 121603 (2013) [arXiv:1307.4768 [hep-th]].
[23] H. A. Gonzalez, J. Matulich, M. Pino and R. Troncoso, JHEP **1309**, 016 (2013) [arXiv:1307.5651 [hep-th]].
[24] A. Campoleoni, S. Fredenhagen, S. Pfenninger and S. Theisen, JHEP **1011**, 007 (2010) [arXiv:1008.4744 [hep-th]].
[25] C. Krishnan, A. Raju and S. Roy, JHEP **1403**, 036 (2014) [arXiv:1312.2941 [hep-th]].
[26] M. Gary, D. Grumiller, M. Riegler and J. Rosseel, JHEP **1501**, 152 (2015) [arXiv:1411.3728 [hep-th]].
[27] G. Barnich, A. Gomberoff and H. A. Gonzalez, Phys. Rev. D **86**, 024020 (2012) [arXiv:1204.3288 [gr-qc]].

[5] In dimensions four and higher these correspond to regions with infinite curvature. In three dimensions the natural adaptation of this idea is an orbifold singularity, like the one we considered in Milne.

[28] B. Burrington, L. A. Pando Zayas and N. Rombes, arXiv:1309.1087 [hep-th].
[29] B. Craps, C. Krishnan and A. Saurabh, JHEP **1408**, 065 (2014) [arXiv:1405.3935 [hep-th]].
[30] C. Krishnan and S. Roy, Phys. Rev. D **88**, 044049 (2013) [arXiv:1305.1277 [hep-th]].
[31] C. Krishnan, A. Raju, S. Roy and S. Thakur, Phys. Rev. D **89**, no. 4, 045007 (2014) [arXiv:1308.6741 [hep-th]].
[32] M. Ammon, M. Gutperle, P. Kraus and E. Perlmutter, JHEP **1110**, 053 (2011) [arXiv:1106.4788 [hep-th]].
[33] A. Castro, E. Hijano, A. Lepage-Jutier and A. Maloney, JHEP **1201**, 031 (2012) [arXiv:1110.4117 [hep-th]].
[34] K. S. Kiran, C. Krishnan, A. Saurabh and J. Simn, JHEP **1412**, 002 (2014) [arXiv:1408.3296 [hep-th]].

Relative Entropy and Torsion Coupling

Feng-Li Lin[1] **and Bo Ning**[2]

[1] Department of Physics, National Taiwan Normal University, Taipei, 11677, Taiwan
[2] Center for Theoretical Physics, College of Physical Science and Technology, Sichuan University, Chengdu, 610064, PR China

E-mail: `fengli.lin@gmail.com`, `ningbbo@gmail.com`

Abstract. Based on the the geometric realization of entanglement entropy via Ryu-Takayanagi formula, in this work we evaluate the relative entropy for the holographic deformed CFT dual to the torsion gravity coupled to the fermions of nonzero vev in the Einstein-Cartan formulation. We find that the positivity and monotonicity of the relative entropy imposes constraint on the strength of axial-current coupling, fermion mass and equation of state. Our work is the first example to demonstrate the nontrivial constraint on the bulk gravity theory from the quantum information inequalities. Especially, this constraint is beyond the symmetry action principle and should be understood as the unitarity constraint. This talk is based on the work [1] of the authors.

1. Introduction

We evaluate the relative entropy on a ball region near the UV fixed point of a holographic conformal field theory deformed by a fermionic operator of nonzero vacuum expectation value. The positivity of the relative entropy considered here is implied by the expected monotonicity of decrease of quantum entanglement under RG flow. The calculations are done in the perturbative framework of Einstein-Cartan gravity in four-dimensional asymptotic anti-de Sitter space with a postulated standard bilinear coupling between axial fermion current and torsion, i.e.,

$$\mathcal{L}_{\psi K} = \frac{\eta_t}{4!}\sqrt{-g}\,\epsilon_{\mu\nu\rho\sigma}\,\bar{\psi}\gamma^{\mu}\gamma_5\psi\,K^{\nu\rho\sigma} \tag{1}$$

where $K^{\nu\rho\sigma}$ is the contorsion tensor. If $\eta_t = 1$, $\mathcal{L}_{\psi K}$ is the canonical coupling of the fermion to torsion as conventionally chosen in [2]. Holographically, this is dual to deforming the holographic CFT by such a coupling term

$$\delta\lambda \int d^d x\, \mathcal{O}_\Delta(x) \tag{2}$$

and with

$$\langle \int d^d x\, \mathcal{O}_\Delta(x)\rangle \neq 0 \tag{3}$$

where $\mathcal{O}_\Delta$ is some single-trace fermion operator of conformal dimension Δ, and $\delta\lambda$ is a spinor characterizes the amount of deformation.

By requiring positivity of relative entropy, our result yields a constraint on axial current-torsion coupling, fermion mass and equation of state. In the following we sketch our results.

IOP Conf. Series: Journal of Physics: Conf. Series **883** (2017) 012016 doi:10.1088/1742-6596/883/1/012016

2. Relative Entropy and Holographic Consideration

The relative entropy is a "distance" measure on the (quantum) state space. For two state ρ and σ, the relative entropy is defined as

$$S(\rho||\sigma) := \mathrm{tr}(\rho \log \rho) - \mathrm{tr}(\rho \log \sigma) \; . \tag{4}$$

As a "distance" measure, the relative entropy is non-negative, i.e.,

$$S(\rho||\sigma) \geq 0 \tag{5}$$

in which the equality holds if and only if $\rho = \sigma$.

For a quantum state one can express the reduced density matrix σ_A on the region A in terms of the modular Hamiltonian H_A as follows:

$$\sigma_A = \frac{e^{-H_A}}{\mathrm{tr}\, e^{-H_A}} \; , \tag{6}$$

then the relative entropy can be rewritten as

$$S(\rho_A||\sigma_A) = \Delta\langle H_A\rangle - \Delta S_A \tag{7}$$

where

$$\Delta\langle H_A\rangle \;:=\; -\mathrm{tr}(\rho_A \log \sigma_A) + \mathrm{tr}(\sigma_A \log \sigma_A) \; , \tag{8}$$

$$\Delta S_A \;:=\; -\mathrm{tr}(\rho_A \log \rho_A) + \mathrm{tr}(\sigma_A \log \sigma_A) \; . \tag{9}$$

Note that the entanglement entropy for state σ on the region A is defined by

$$S_A(\sigma) := -\mathrm{tr}(\sigma_A \log \sigma_A) \tag{10}$$

so that ΔS_A is the difference of entanglement entropy of the region A between states ρ and σ. Via (7) the positivity of relative entropy yields

$$\Delta\langle H_A\rangle \geq \Delta S_A \; . \tag{11}$$

The modular Hamiltonian H_A is in general nonlocal and unknown. However, for CFT$_d$ with the interested region A to be a disk of radius R_A, it has a closed form as follows [3]:

$$H_A = 2\pi \int_{|x|<R_A} d^{d-1}x \, \frac{R_A^2 - r^2}{2R_A} \, T_{tt}(\vec{x}) \tag{12}$$

where T_{tt} is the (holographic) energy density operator of CFT.

To extract some constraints from the relative entropy, we are interested in the case where σ is the CFT vacuum state, and ρ is a perturbative state (by either excitation or deformation) away from σ, i.e.,

$$\rho = \sigma + \delta\rho \; , \qquad |\delta\rho|/|\rho| << 1 \; . \tag{13}$$

Since the relative entropy takes its extremum at $\rho = \sigma$, its first order variation vanishes, which then yields the first law of entanglement thermodynamics [4], i.e.,

$$\Delta\langle H_A\rangle|_{\mathcal{O}(\delta\rho)} = \Delta S_A|_{\mathcal{O}(\delta\rho)} \; . \tag{14}$$

On the other hand, the positivity of the second order variation will impose some unitarity bound on the deformed states of CFT.

IOP Conf. Series: Journal of Physics: Conf. Series **883** (2017) 012016 doi:10.1088/1742-6596/883/1/012016

In the context of AdS/CFT correspondence, a dual state is characterized by an asymptotic AdS bulk metric, i.e., in the Poincare coordinates,

$$ds^2 = \frac{\ell^2}{z^2}\Big(G(z)dz^2 + H_{\mu\nu}(z, x^\mu)dx^\mu dx^\nu\Big) \tag{15}$$

where ℓ is the AdS radius. The CFT vacuum state σ corresponds to $G(z) = 1$ and $H_{\mu\nu} = \eta_{\mu\nu}$. The perturbative state ρ will be given by some slightly deviated $H_{\mu\nu}$ as well as $G(z)$.

We can then obtain the modular Hamiltonian H_A on the region A from holographic $T_{\mu\nu}$ for the corresponding state by the relation (12) if A is a ball. By evaluating H_A for both ρ and σ states and subtracting them, we can then obtain $\Delta\langle H_A\rangle$. We still need to calculate ΔS_A holographically in order to obtain the relative entropy by (7). The entanglement entropy can be obtained holographically by the Ryu-Takayanagi formula [5]:

$$S_A = \frac{\text{Area}(\gamma_A)}{4G_N} \tag{16}$$

where γ_A is the co-dimensional two extremal surface ending on the entangling surface ∂_A at the AdS boundary $z = 0$. For static metric, γ_A is simply the minimal surface on a fixed time slice. We can then evaluate the γ_A and thus S_A for the bulk metrics corresponding to states ρ and σ, and then subtract them to obtain ΔS_A.

3. Summary of Our Results

In this work we consider the deformed holographic CFT with fermionic sources in AdS_4 space in the framework of Einstein-Cartan gravity with a postulated coupling constant for the bilinear interaction between axial current and torsion. Note that this coupling cannot be fixed by the dynamical symmetry of holographic CFT. Instead we find that the positivity of relative entropy disfavors such a coupling.

By turning on both normalizable and non-normalizable fermionic zero modes, the bulk geometry will be backreacted by the torsion at the second order of gravitational coupling. We will solve the backreacted metric in the expansion of Newton constant, denoted by κ, up to κ^2 order, i.e.,

$$\mathbf{g} = \mathbf{g}_0 + \kappa\, \mathbf{g}_1 + \kappa^2\, \mathbf{g}_2 \tag{17}$$

where $\mathbf{g}_0$ is the pure AdS_4 metric. Note that $\mathbf{g}_1$ and $\mathbf{g}_2$ will encode the information about the fermionic sources.

Moreover, a bulk fermion of mass m considered here is dual to a CFT operator of conformal dimension Δ given by

$$\Delta = \frac{d}{2} + |m|\,. \tag{18}$$

Unlike the case for the scalar field/operator, from (18) we see no analogue of BF bound on m for preventing from instability.

In this paper, we will consider the backreaction due to the non-vanishing stress tensor of the bulk fermionic zero modes. This is dual to the holographic deformed CFT given by (2) and (3). Then, we will evaluate the relative entropy up to κ^2 (or δl^2) order by using the backreacted metric.

As we will see, the torsion will affect $\mathbf{g}_2$ but not $\mathbf{g}_1$. Thus, at the κ order, everything should go as the usual Einstein gravity so that the relative entropy evaluated holographically will satisfy the first law (14), i.e.,

$$S(\rho||\sigma)|_{\text{K}} = \Delta\langle H_A\rangle|_{\text{K}} - \Delta S_A|_{\text{K}} = 0\,, \tag{19}$$

as expected.

For the evaluation of the second order relative entropy, we first need to evaluate $\Delta\langle H_A\rangle|_{\mathrm{K}^2}$ by plugging $\langle T_{tt}\rangle|_{\mathrm{K}^2}$ into (12). However, the holographic evaluation of $\langle T_{tt}\rangle|_{\mathrm{K}^2}$ via $\mathbf{g}_2$ yields zero. Thus, we have $\Delta\langle H_A\rangle|_{\mathrm{K}^2} = 0$. Combined this result with the one of (19), we obtain the relative entropy (7) up to K^2 order, that is

$$S(\rho||\sigma) = -\Delta S_A|_{\mathrm{K}^2} \geq 0\,. \tag{20}$$

To evaluate $\Delta S_A|_{\mathrm{K}^2}$, we need to first evaluate the the minimal surface γ_A with respect to the metric $\mathbf{g}_0 + \mathrm{K}\mathbf{g}_1$ up to K^2 order, and we denote it by $\gamma_A^{(1)}$. Then, we can obtain $\Delta S_A|_{\mathrm{K}^2}$ by

$$\Delta S_A|_{\mathrm{K}^2} := \frac{\mathrm{Area}(\gamma_A^{(1)})|_{\mathbf{g}}}{4G_N}|_{\mathrm{K}^2} \tag{21}$$

where $\mathrm{Area}(\gamma_A^{(1)})|_{\mathbf{g}}$ means evaluating the area of surface $\gamma_A^{(1)}$ with respect to the metric (17).

In this work we find that the constraint (20) yields a constraint on m:

$$m^2\ell^2 \geq \frac{2\eta_t^2}{\mu_0^2} \tag{22}$$

where η_t is the postulated coupling constant for the bilinear interaction between bulk axial current and torsion, and μ_0 characterizes the equation of state for the dual deformed fermion state by

$$P = \frac{\mu_0 - 2}{2\mu_0}\,\varepsilon\,. \tag{23}$$

Here P is the pressure and ε is the energy density of the deformed fermion state. Note that the monotonicity condition $\frac{\partial S(\rho||\sigma)}{\partial R_A} \geq 0$ yields the same condition (22).

We see that the positivity of the relative entropy imposes constraint on the bilinear coupling η_t in (1) as well as the fermion mass m and equation of state μ_0. If there were no such coupling, i.e, $\eta_t = 0$, then the relative entropy is positive for all m, which is consistent with the fact there is no BF-bound for fermion's mass and seems more natural. Otherwise, the constraint is on all three parameters in a nontrivial way. It is interesting to see how this constraint can be understood as some energy condition in the bulk Einstein-Cartan gravity.

Acknowledgements
FLL would like to thank APCTP for the travel support to attend and present this work in IF-YITP GR+HEP+Cosmo International Symposium VI held at IF, Naresuan University, Thailand, 3-5 August, 2016. This work is supported by Taiwan Ministry of Science and Technology through Grant No. 103-2112-M-003 -001 -MY3 and 103-2811-M-003 -024.

References
[1] B. Ning and F. L. Lin, "Relative Entropy and Torsion Coupling," Phys. Rev. D **94**, no. 12, 126007 (2016) [arXiv:1607.00263 [hep-th]].
[2] F. W. Hehl, P. Von Der Heyde, G. D. Kerlick and J. M. Nester, "General Relativity with Spin and Torsion: Foundations and Prospects," Rev. Mod. Phys. **48**, 393 (1976).
[3] H. Casini, M. Huerta and R. C. Myers, "Towards a derivation of holographic entanglement entropy," JHEP **1105**, 036 (2011). [arXiv:1102.0440 [hep-th]].
[4] T. Faulkner, M. Guica, T. Hartman, R. C. Myers and M. Van Raamsdonk, "Gravitation from Entanglement in Holographic CFTs," JHEP **1403**, 051 (2014). [arXiv:1312.7856 [hep-th]].
[5] S. Ryu and T. Takayanagi, "Holographic derivation of entanglement entropy from AdS/CFT," Phys. Rev. Lett. **96**, 181602 (2006). [hep-th/0603001].